The Death of the Comprehensive High School?

SECONDARY EDUCATION IN A CHANGING WORLD

Series editors: Barry M. Franklin and Gary McCulloch

Published by Palgrave Macmillan:

The Comprehensive Public High School: Historical Perspectives
By Geoffrey Sherington and Craig Campbell
(2006)

Cyril Norwood and the Ideal of Secondary Education
By Gary McCulloch
(2007)

*The Death of the Comprehensive High School?:
Historical, Contemporary, and Comparative Perspectives*
Edited by Barry M. Franklin and Gary McCulloch
(2007)

The Death of the Comprehensive High School?

Historical, Contemporary, and Comparative Perspectives

Edited by

Barry M. Franklin
and Gary McCulloch

palgrave
macmillan

THE DEATH OF THE COMPREHENSIVE HIGH SCHOOL?
Copyright © Barry M. Franklin and Gary McCulloch, 2007.

All rights reserved. No part of this book may be used or reproduced in any manner whatsoever without written permission except in the case of brief quotations embodied in critical articles or reviews.

First published in 2007 by
PALGRAVE MACMILLAN™
175 Fifth Avenue, New York, N.Y. 10010 and
Houndmills, Basingstoke, Hampshire, England RG21 6XS
Companies and representatives throughout the world.

PALGRAVE MACMILLAN is the global academic imprint of the Palgrave Macmillan division of St. Martin's Press, LLC and of Palgrave Macmillan Ltd. Macmillan® is a registered trademark in the United States, United Kingdom and other countries. Palgrave is a registered trademark in the European Union and other countries.

ISBN-13: 978–1–4039–7769–4
ISBN-10: 1–4039–7769–0

Library of Congress Cataloging-in-Publication Data

    The death of the comprehensive high school? : historical, contemporary, and comparative perspectives / Barry M. Franklin and Gary McCulloch, editors.
        p. cm.—(Secondary education in a changing world)
    Includes bibliographical references and index.
    ISBN 1–4039–7769–0 (alk. paper)
      1. Comprehensive high schools—Cross-cultural studies. 2. Education, Secondary—Cross-cultural studies. I. Franklin, Barry M. II. McCulloch, Gary. III. Series.

LB43.D43 2007
373.250973—dc22   2007010723

A catalogue record for this book is available from the British Library.

Design by Newgen Imaging Systems (P) Ltd., Chennai, India.

First edition: December 2007

10 9 8 7 6 5 4 3 2 1

Printed in the United States of America.

Contents

Introduction

1 Introduction—The Death of the Comprehensive
  High School? Historical, Contemporary, and
  Comparative Perspectives   3
  Barry M. Franklin and Gary McCulloch

Historical Perspectives

2 Custodialism and Career Preparation in a
  Comprehensive High School, 1929–1942   19
  Sevan G. Terzian

3 *Education for All American Youth* (1944): A Failed
  Attempt to Extend the Comprehensive High School   37
  Wayne J. Urban

4 The Comprehensive High School, Enrollment
  Expansion, and Inequality: The United States
  in the Postwar Era   53
  John L. Rury

Contemporary Perspectives

5 Breathing Life into Small School Reform: Advocating
  for Critical Care in Small Schools of Color   73
  Rene Antrop González and Anthony De Jesús

6 Soul Making in the Comprehensive High School:
  The Legacies of Frederick Wiseman's *High School*
  and *High School II*   93
  José R. Rosario

7 The End of the Comprehensive High School?
  African American Support for Private School Vouchers   111
  Thomas C. Pedroni

Comparative Perspectives

8 The Formation of Comprehensive Education:
   Scandinavian Variations 131
   Susanne Wiborg

9 Missing, Presumed Dead? What Happened to the
   Comprehensive School in England and Wales? 147
   David Crook

10 The Comprehensive Ideal in New Zealand:
   Challenges and Prospects 169
   Gregory Lee, Howard Lee, and Roger Openshaw

11 "My Parents Came Here with Nothing and
   They Wanted Us to Achieve": Italian
   Australians and School Success 185
   Pavla Miller

Epilogue

12 Epilogue—The Future of the Comprehensive
   High School 201
   Gary McCulloch and Barry M. Franklin

Contributors 205

Index 207

Introduction

Chapter 1

Introduction—The Death of the Comprehensive High School?
Historical, Contemporary, and Comparative Perspectives

Barry M. Franklin and Gary McCulloch

The comprehensive high school has become familiar in different national contexts around the world as a means of providing for the education of adolescents. Its prime rationale is to educate pupils of all abilities and aptitudes within the same educational institution in order to provide equality of opportunity for all. For many years this basic model, drawing in different ways on the prototype developed in the United States, seemed to be extending its reach to different societies around the world, becoming firmly established and often dominant, and eclipsing alternative approaches that have usually been based on some form of selection, for example, by parental ability to pay for separate schooling or by competitive examination for different kinds of schools.

However, over the past two decades, and especially in the early years of the twenty-first century, the comprehensive high school has encountered a number of major challenges that have caused it to come under severe scrutiny. Indeed, when educators, politicians, policy makers, and even ordinary citizens in any of a number of industrialized countries talk about the high school, they more often than not describe it as troubled or as the most troubled of our educational institutions. The most important of these problems lies with the question of who the high school should serve. Should this institution serve all adolescents or should it be one of many institutions that serve some youth and not others? A related challenge has to do with the question of what the high school should offer. Should it be an agency that provides a common educational experience for all adolescents or should it be a specialized institution that provides for the unique needs of particular youth?

A third challenge has to do with the purpose of this institution. Is the high school the setting for preparing adolescents with the knowledge, skills, and dispositions required of a citizen in a democratic society? Or is it the site for training the nation's workforce? Or is the high school the agency that secures the social mobility of adolescents? Although these goals are often contradictory, those enamored of the principle of comprehensiveness have argued that the high school should attempt to fulfill all of these purposes at one and the same time (Achieve, 2005; American Diploma Project, 2005; Cohen, 2001). It is these issues that the essays contained in this volume explore.

This edited volume investigates the nature and extent of these challenges in major westernized societies. It will do so by addressing two fundamental questions about the comprehensive high school. The first question is whether we are witnessing the death of the comprehensive high school in the principal westernized societies in which it is located. Alternatively, are reports of its demise exaggerated? This volume addresses this question by looking in depth at the institution in the land of its birth, the United States. Essays examine any symptoms of the decline, dispersal, or decay of this institution and the model of schooling that it represents, in some cases by appraising tensions and contradictions that may have developed over time even during the period of its emergence and hegemony. Similarly, the volume includes case studies that address these same questions in other westernized societies, specifically Britain, Scandinavia, Australia, and New Zealand.

The volume also seeks to pursue these themes by posing a second key question. To the extent that the comprehensive high school is in difficulties and potential or actual decline, what reasons can be advanced for this? Can its problems be explained by local factors that differ from one setting to another? Are these problems symptomatic of global factors affecting all comprehensive high schools, regardless of their location? Or are these problems the result of a more complex combination of local, national, and global factors, with effects that vary across similar and different settings?

In addressing these aspects, we seek to highlight the relationship between historical and contemporary developments. The comprehensive high school is itself a historical phenomenon, with roots in the nineteenth and twentieth centuries rather than in the twenty-first. This history is part of the longer-term development of secondary education as a whole, whose lineage stretches back in time over many centuries (see, e.g., Durkheim, 1977; Savoie, Bruter, and Frijhoff, 2004; McCulloch, 2007). To assess its present condition and its future prospects, we must come to terms with its past. Even within its own terms, the history of the comprehensive high school is substantial enough for us to be able to analyze its changing fortunes in different national contexts in some detail. This helps us in turn to understand what Campbell and Sherington in their history of the comprehensive public high school in New South Wales, Australia, characterize as "the origin of the present discontents" (Campbell and Sherington, 2006, p. 1).

The comparative and international dimensions of the comprehensive high school constitute a further significant theme of the present volume. Accounts of these schools have hitherto tended to focus on particular national circumstances, largely in isolation from those prevailing elsewhere. In the past decade, however, there have been signs of growing interest in international patterns of development in official policy circles. This has largely been a response to anxieties about secondary education, and part of a search for alternative possibilities. At one level, such policy initiatives have arisen from national debates. In Britain, for example, the Select Committee on Education and Skills, reporting to Parliament, undertook an extensive inquiry into secondary education, beginning in 2001, that included visits to the cities of Birmingham in England and Auckland in New Zealand, "to get to grips with the issues facing secondary education and to look at different approaches to them" (Education and Skills Committee, 2003, Introduction).

In another respect, international organizations have made a notable contribution to this new development. Thus, the Organization for Economic Co-operation and Development (OECD) has initiated the Program for International Student Assessment (PISA) as a three-yearly survey, beginning in 2000, of the knowledge and skills of 15-year-olds in industrialized countries (see OECD, 2003). This has led to comparisons being drawn between different countries and cultures. Green and Wiborg, for example, have used data provided by OECD surveys to develop a fuller estimation of the effects of different school systems on educational inequality across countries. They conclude on this basis that the more equal countries have in common "the structures and processes typically associated with radical versions of comprehensive education: non-selective schools, mixed ability classes, late subject specialization and measures to equalize resources between schools" (Green and Wiborg 2004, pp. 239–40). The World Bank has also provided commentary and analysis as well as funding for secondary education. Interestingly, a major World Bank report on secondary education refers to a "U.S. template" for the expansion of secondary education, distinctive from the "European pattern of secondary schooling." It suggests that this template encompasses public funding and provision, a nonselective system with no early specialization or academic segregation, an academic yet practical curriculum, numerous small and fiscally independent school districts, and secular control of schools and school funds. However, it argues, critics now see these features as tending to constrain rather than advance education, and this critical assessment is at the core of the ongoing debate in the further development of secondary education around the world (World Bank, 2005, p. 3). The present volume helps to assess the extent of the continuing vitality of the "U.S. template" and the nature of the pressures that surround it, both within the United States and beyond.

Another key concern of the current volume is to promote the use of a wide range of sources and methods in the critical study of comprehensive high schools. Against a backcloth of historical, contemporary, and comparative issues, we can develop a number of research strategies to enhance our understanding of this multifaceted phenomenon. Included here are the quantitative analysis of enrollment expansion and inequality, case studies of individual schools and districts, study of specific reports in relation to their broader context, appraisal of visual evidence, and the use of a wide range of documentary sources to gauge the changing social and political relationships of the comprehensive high schools. Our intention is to encourage further research with these broad perspectives that embrace a wide range of evidence and source material, and thus to challenge the insularity, presentism, and fixation on single-method approaches that are currently prevalent in much educational research (see also, e.g., McCulloch and Richardson, 2000; McCulloch, 2004).

Historical Perspectives

In the United States the mid-nineteenth century high school was a selective institution in which only a small minority of children ever attended. These were children whose family resources allowed them to attend schooling beyond the elementary grades, whose future lay not in the professions and in attending college but in middle management positions in the nation's expanding commercial and manufacturing sectors, and whose abilities allowed them to pass a rigorous entrance examination (Labaree, 1988; Reese, 1995).

By the end of the nineteenth century, admission examinations had disappeared from most of the nation's high schools. Coupled with growing enrollments that continued through the first half of the twentieth century, the high school had become by the end of World War II a mass institution. Dubbed the comprehensive high school to acknowledge its broad purposes and extensive programs, it has come to enroll a diverse array of students, some preparing for college, some for immediate entry into the workforce, and others whose future was at best uncertain. Central to its wide mission has been the existence of a differentiated curriculum composed of distinct programs, some academic and preparatory and others vocational and terminal. Such an array of offerings has enabled the high school to be accessible to all youth. Once inside the schoolhouse door, however, differentiation has made it possible for high schools to sort students into an array of different courses of study and ultimately into very different and unequal educational and life destinies (Conant, 1959, 1967; Franklin, 1986; Kliebard, 2004; Krug, 1964, 1972).

For some, the wide accessibility of the comprehensive high school has made it the preeminent instrument for advancing equal opportunity and ultimately American democracy itself (Tanner and Tanner, 1990; Wraga, 1994). Yet, others have been critical of this attempt to serve everyone. They argue that the effort of high school leaders to provide for all students has led them to establish a multitude of programs built on an array of elective courses that ultimately has produced a fragmented curriculum that lacks a unified purpose and focus. They also argue that the attempt of high school leaders to create an institution that is responsive to the particular demands of virtually every student has compromised academic rigor and demanding standards. The relationship between teachers and students has become something akin to a treaty or bargain characterized by minimal demands in return for minimal compliance and effort. As a result, it is often suggested, the effort of the high school to provide for everyone has created an institution that serves no one particularly well (Angus and Mirel, 1999; Grant, 1988; Powell, Farrar, and Cohen, 1985; McNeil, 1986, 2000; Sedlak, Wheeler, Pullin, and Cusick, 1986).

The differing purposes that early twentieth-century school reformers set for the high school has been described as a battle between academic and career preparation on the one hand and custodialism on the other. Sevan Terzian's case study (chapter 2) of Ithaca High School at the end of the decade of the 1920's through the beginning of the 1940s explores this conflict. Some historians of the American high school have pointed to the Great Depression and World War II years as a pivotal period when academic and vocational preparation gave way to a new custodial mission characterized by the emergence of less rigorous and terminal academic subjects. Others have hinted at a different meaning of custodialism in the comprehensive high school: as a kind of civic preparation for democratic life through the extracurriculum. This chapter considers the perspectives and priorities of school administrators and students during the Great Depression and at the onset of World War II at one comprehensive high school—Ithaca High School (IHS) in Ithaca, New York—to determine the extent to which academic and civic custodialism appeared during a period of significant institutional change. Utilizing district annual reports, superintendent bulletins, board of education meeting minutes, guidance department follow-up surveys, and student newspapers, it concludes that the administration intended to further its vocational and academic preparation of students. Many former students, however, complained that Ithaca High School had failed to prepare them adequately for work or higher education. Some students writing in the school newspaper, meanwhile, complained that the school's poorly supported extracurricular activities did not prepare them for their future roles as citizens. These multiple perspectives and mixed assessments suggest that some aspects of academic custodialism began to appear at IHS and that civic custodialism was not very much in evidence. The chapter concludes by proposing that this case history can remind current critics and policy makers that effective education reform must not neglect the civic dimension of the comprehensive high school as well as the perspectives of students and parents.

High school reform has had a long history dating back to the last decade of the nineteenth century. A host of national committees including but not limited to the National Education Association's Committee of Ten in 1894, the Commission on the Reorganization of Secondary Education in 1918, the Progressive Education Association's Eight Year Study beginning in 1932, the reports of the U.S. Office of Education on life adjustment education during the late 1940s, and work of the National Commission on Excellence in Education in 1983 have, each in their own way, offered a remedy for the problems facing the high school (Kliebard, 2004; Ravitch, 2000). Wayne Urban's essay (chapter 3) considers one of the most important of these high school reform initiatives, *Education for All American Youth*. A product of the deliberations of the Educational Policies Commission (EPC) during the early 1940s, the report, which was published in 1944, represented the attempt of the American educational establishment to extend the reach of the comprehensive high school through what would be called grades 13 and 14. Focusing on vocational training and guidance, the report represented, at least for the EPC, how professional educators of the day would reform the American high school.

Invoking the language of educational progressivism and avoiding discussions about the teaching of academic content, the EPC report attempts to capture more adolescents for the public schools by incorporating grades 13 and 14, which were in effect the first two years of college, into the high school. Ironically, as Urban shows, the publication of the famous Harvard Redbook, *General Education in a Free Society* (1945), robbed the EPC of center stage in the discussions of secondary education and aborted its campaign to expand the comprehensive high school before it could get off the ground. Urban's essay offers a good picture of how efforts to reform the American high school are short-lived and rarely get anywhere. It is a sober reminder of what might happen to the ambitious plans of today's high school reformers as they confront the real world of educational policy making.

The post–World War II period in the United States was a time of rapid expansion for the American high school. This growth and its interplay with issues of inequality is the theme of John Rury's essay (chapter 4). This chapter examines the growth of secondary enrollments in the United States between 1950 and 1970, a time of rapid educational expansion around the world. Major theories of educational expansion are considered. Specifically, the influence of economic and technological factors on enrollment growth is assessed, along with diffusion models that hold that education expanded of its own momentum during this period. These frameworks of analysis are examined with the use of state-level census data from 1950 and 1970. Utilizing multiple regression, the study finds that the diffusion perspective is most consistent with the growth of secondary schooling in the United States during this time. In other words, the growth of high school enrollments does not appear to have been linked to changes in the labor force often associated with the emergence of a white collar or "knowledge" economy. The analysis also points to strong regional differences in secondary schooling, which appear to close substantially during the postwar era. At least some of this change appears to have been due to growth in black high school enrollments. These findings provide caveats to the standard diffusion model discussed in the sociological literature. The study also examines variation in metropolitan secondary enrollments in 1970 to take stock of high schools at the end of the period in question. Variation is somewhat less than it had been in 1950, but poverty levels in these urban areas bear a strong and significantly negative association with secondary enrollment levels. Southern metropolises continue to lag behind those elsewhere, suggesting that important regional differences persisted despite convergence in the preceding two decades. Part-time youth employment and white collar employment also were associated with enrollment levels, suggesting that economic factors continued to impact secondary schooling, even if they were unrelated to educational expansion. By the end of this period, despite impressive growth in secondary participation, considerable inequality continued to characterize schooling in the United States. This, along with dissatisfaction over curricular issues, would lead reformers to look beyond the comprehensive high schools in the decades following the 1970s.

Contemporary Perspectives

In 2005, the National Governors Association and Achieve, Inc., a partnership of the Association and a number of key national business leaders, devoted its annual Education Summit to the issue of high school reform. The impetus for the summit was the belief that the contemporary high school has failed to prepare students for either higher education or employment. Pointing to the poor showing of the United States in international comparisons of mathematics and science achievement and the high school's low graduation rates, the Summit called for an array of reforms. They advocated the introduction of a more rigorous and challenging curriculum, the enhancement of the training provided to high school teachers and principals, the introduction of content standards that would hold high schools accountable for student achievement, the development of governance mechanisms that would allow better coordination and cooperation between high schools and postsecondary institutions, and the redesign of the high school. As the governors saw it, many of the nation's comprehensive high schools were "too impersonal, inflexible, and alienating" for many adolescents. What was called for in the way of redesign was the breaking up of high schools into smaller learning communities and the creation of new small high schools (Achieve, Inc., 2005, p. 9). The National Governors Association was not the only body calling for the creation of smaller high schools. The Aspen Institute (Cohen, 2001) and the Bill and Melinda Gates Foundation (2005) have also called for the creation of smaller high schools. So too have a number of contemporary educational researchers (Fine, 2005; Meier, 1995, 2002; Toch, 2003).

In their essay, Rene Antrop-Gonzalez and Anthony De Jesús (chapter 5) offer a case study of two such smaller schools. Using in-depth interviewing, participant observations, and the review of historical and curricular documents, their chapter describes and analyzes two Latino community-based small high schools—the Dr. Pedro Albizu Campos High School (PACHS) in Chicago and El Puente Academy for Peace and Justice (El Puente) in New York City. They suggest that unlike failing comprehensive high schools located in urban areas, these schools are successful because they foment a culture of high academic expectations for their students, value high-quality interpersonal relationships between students and teachers, and privilege the funds of knowledge that students and their respective communities bring to school. Based on these findings, they spell out a theory of critical care that embodies these necessary conditions if small high schools created for and by communities of color are to succeed.

Key to the potential success that small schools offer the task of school reform is their ability to change the regulative mechanism of the school from an overt form of policing to a more indirect form of community building. This is in effect a shift from overt and explicit modes of control to more indirect and implicit forms. In his chapter, Jose Rosario (chapter 6) explores this transition through an analysis of two documentary films that examine the role of the high school in governing behavior. The first film, *High School*, was produced by the documentary film maker Frederic Wiseman in 1968 and looks at a large—almost 4,000 student—comprehensive, predominantly white, middle-class high school in Philadelphia, Pennsylvania. The second film, *High School II*, was filmed by Wiseman in 1992 and examined life at New York City's Central Park East Secondary School, a largely black school in Central Harlem with an enrollment of about 480 students. For our purposes in this volume, these two schools point to the changes that are occurring throughout the United States in the organization of the high school.

Issues of access, however, have not been absent within this supposedly comprehensive setting. The locations of high schools within communities and the practice of assigning students to such schools on the basis of their place of residence introduces a question about the openness of such schools to various segments of the population. The secondary schools of any given community often serve students of distinct class and ethnic and racial backgrounds with different occupational and social aspirations. The historic and ongoing existence of racial segregation is the most blatant example of this form of selection. The fact that such schools have historically had different and unequal resources has exacerbated this differentiation among schools.

Within secondary schools themselves there has been another form of selection through the practice of curriculum differentiation. Here, students are sorted by ability, which is often a surrogate for race and class, and then channeled into different and unequal educational programs that themselves lead to different and unequal occupational and social roles in adult society. These forms of selection have over the years and in different places waxed and waned in response to the influence and impact of different educational philosophies, local conditions, national priorities, and other social forces. There has of course been another way in which issues of access have affected the ability of secondary schools to ensure equality and promote academic achievement. Private secondary schools, religious and secular, have always served to divide students not only by parental income and faith traditions but by race and social class.

In his essay, Thomas Pedroni (chapter 7) suggests that our assumptions about the divisive nature of private schools may be too simplistic. Looking at Milwaukee's Parental Choice Program, Pedroni asks whether voucher programs signal the death of support among many African American families for the comprehensive high school. He also asks whether African American voucher parents have lost interest in the promise of the comprehensive high school, or do they herald vouchers as the vehicle that might finally deliver some of the more valued elements of that promise. Pedroni's essay examines the points of overlaps and departures among the educational visions of three sets of stakeholder within the city's voucher debate: African American voucher families, the more powerful social forces that fund market-based reform, and defenders of the comprehensive school vision.

Pedroni argues that rather than producing greater variety and market segmentation in educational forms—a direct contradiction of the universalist ethos of the comprehensive high school—much research points to the conclusion that markets actually produce more, not less educational standardization. Therefore, free markets are not essentially counterproductive to furthering the more universalist comprehensive high school. In fact, many private schools actually better approximate the comprehensive high school than urban public high schools, the latter well known to working-class families of color for their tremendous homogeneity in both demographic makeup and academic offerings.

Given this, the task for supporters of the public comprehensive high school is to recognize that African American voucher families are not rejecting the comprehensive high school but rather a public high school which never attained anything like the stature that promoters of the comprehensive high school envisioned. Public urban high schools intended for working-class people of color have never deviated much from being overwhelmingly segregated institutions offering a singular curriculum focused on the disciplinary control of the population contained in their buildings. Although there is an implicit critique of the comprehensive high school built unto the narratives of the parents that Pedroni interviewed—that the comprehensive high school does not offer the more intimate school size, sense of community, and individualized attention that many voucher parents crave—nevertheless parents actually yearned for greater heterogeneity among both students and curriculum tracks than the public schools had to offer. That is, they were rejecting some of the same qualities in the public schools that comprehensive school advocates would reject.

Comparative Perspectives

By contrast with the United States, secondary education in Europe has usually been regarded as notably differentiated and segregated in nature. In Western Europe the key position was held by schools with a strong academic tradition, recruiting pupils by selective methods and in turn closely linked with universities that were relatively restricted in numbers. The French *lycee*, the German *gymnasium*, and the English grammar school represented exemplars of this tendency. These institutions appeared to be less well adapted than was the American model of the comprehensive high school to the demands of a modern industrial society. Again unlike their American counterpart, they served to block popular demands for the expansion of secondary education.

Notwithstanding this dominant trend, there have been regions of Europe, for example Scandinavia, that have promoted a comprehensive rather than a selective model of secondary education. In her essay, Susanne Wiborg (chapter 8) offers a comparative explanation as to why the Scandinavian countries developed a radical system of comprehensive education. The essay seeks to uncover a single set of factors that the Scandinavian countries share in common—and that are absent in other countries—in order to develop a general explanation of why these countries went along the same track when introducing comprehensive education. Scandinavian school history has fostered numerous single-case studies that in outline in detail the specific national development of comprehensive education. However, even though these studies have provided us with important contextualized accounts, they are nevertheless so deeply entrenched in national history that generalization beyond the particular is almost impossible. In order to explain why the Scandinavian countries have introduced radical comprehensive education simply requires a search beyond the particular for a single set of factors that determines this common outcome. It will be argued that the overall comparative explanation has mainly to do with the unique political tradition of consensus seeking politics between the Liberal and Social Democratic parties. The making of the peasantry into an independent class that subsequently constituted the Liberal Party with socially liberal views strong enough to crush the Right, and the rise of a Social Democratic Party that welded an alliance with the Liberals goes far in explaining how a radical tradition of education could be introduced through broad coalitions.

Over time, this European tradition of selective secondary education has gradually given way to something more akin to the American system of comprehensive secondary education. On the other hand, as has been the case in the United States, new social and political movements favoring an emphasis on parental choice and diversity of provision have in many contexts led to a renewed emphasis on differentiation within and between schools. Rather than a convergence of national systems toward comprehensive ideals, it might instead be argued that there is in the early twenty-first century a great deal of pressure in the other direction, toward a reinforcement of selection and social difference.

David Crook's essay (chapter 9) offers an illustration of the interplay between comprehensiveness and selection in England and Wales from the 1920s onward. Crook notes that today most English secondary schools have a specialist designation. End-of-driveway entrance signs including the words "comprehensive school" have been vanishing. In view of this, and of Tony Blair's 2002 announcement of the need to move to the "post-comprehensive era," it is important to discover and explain what happened to the comprehensive school in England and Wales. The essay traces its origins back to the 1920s but focuses attention on the post–World War II period and the decades of the 1960s and 1970s. It was in this latter period that successive governments disagreed as to whether comprehensive schooling should be national policy, even though there was no common understanding of what a comprehensive school is or should be.

The essay explores how the process of "going comprehensive" has been both uneven and incomplete. Crook examines a range of historical representations and opinions relating to the history of comprehensive schooling raising questions about truth and bias, success and failures. Applying analogies suggested by the use of the term "death" in the title of this book, Crook suggests that the comprehensive school enjoyed better health in Wales than in England, where opinions differed about its efficacy. Having been neglected since the 1980s, the comprehensive school is missing, presumed dead, though its distant cousin, comprehensive education, remains alive.

The battleground for the future of secondary education is also being fought increasingly on a global scale, as in different parts of the world the mass migration of peoples and new forms of communication have a major effect on debates and practices. This is evident, for example, in New Zealand and Australia, where recent research has shown that regional factors and changing educational markets are affecting the character of comprehensive schools, and playing a systematic role in the production of differentiated school outcomes. In their essay, Gregory Lee, Howard Lee, and Roger Openshaw (chapter 10) trace the development of a comprehensive postprimary schooling model in New Zealand throughout the twentieth century. They begin with a discussion of the introduction of "free place" postprimary education in 1901, trace the gradual and at times controversial translation of rural (district high) and urban (technical high) postprimary schools into comprehensive secondary institutions by the mid-1970s, and conclude with a discussion of the appropriateness of a comprehensive secondary education model for New Zealand in the early twenty-first century.

During this period an extensive compulsory common core curriculum was developed and incorporated into the three existing types of postprimary schools: district high schools, secondary schools, and technical high schools. Not surprisingly perhaps, curricular overlap between these institutions—already the subject of considerable debate—became even more pronounced with the launching of a general education curriculum in 1946 which, in turn, sought to counter conservative arguments for retaining separate secondary and technical high schools in particular. They conclude arguing that the adoption of a comprehensive postprimary curriculum in postwar New Zealand signaled to potential critics the end of what had long been regarded as a rigid and highly selective tripartite schooling model.

Taken together, the essays that comprise this book up to this point suggest an uncertain future for the comprehensive high school. From their outset, as the historical essays in this book suggest, the comprehensive high school has had difficulty in attaining the broad mission that its founders had set for it. The essays in this volume that consider more contemporary issues and events also point to this as a continuing dilemma. In the last essay in this volume, Pavla Miller (chapter 11) looks to Australia to indicate why this institution has such an uncertain future. In Australia, according to Miller, the unfinished project of making high schools comprehensive is under attack. Today, the country has one of the strongest private educational sectors among OECD countries. There is compelling evidence that such a strong private sector is contributing to social polarization. In her essay, Miller considers two recent studies that address this process of privatization. One of the studies employs social theory and survey data to explore how the educational markets that privatization brings compound social polarization. The second study explores the history of the comprehensive high school in one Australian state and uses the experiences of three generations of Italian-Australians in the City of Melbourne to explore the impact of privatization. The book concludes with an epilogue (chapter 12) in which the volume's editors examine the implications of these studies for our understanding of the current state and future prospects of the comprehensive high school.

References

Achieve, Inc. (2005). *An action agenda for improving America's high schools*. Washington: D.C.: National Governors' Association.
Angus D.L. and Mirel, J. (1999). *The failed promise of the American high school 1890–1995*. New York: Teachers College Press.
Bill and Melinda Gates Foundation (2005). *High school for the new millennium*. Seattle: Bill and Melinda Gates Foundation.
Campbell, C. and Sherington, G. (2006). *The comprehensive public high school: Historical perspectives*. New York: Palgrave Macmillan.
Cohen M. (2001). *Transforming the American high school. New directions for state and local policy*. Washington, D.C.: The Aspen Institute.
Conant, J.B. (1959). *The American high school today: A first report to interested citizens*. New York: McGraw-Hill Book Company.
———. (1967). The *comprehensive high school: A second report to interested citizens*. New York: McGraw-Hill Book Company.
Durkheim, E. (1977). *The evolution of educational thought: Lectures on the formation and development of secondary education in France*. London: Routledge and Kegan Paul.
Education and Skills Committee. (2003) *Second Report*. London: United Kingdom Parliament.
Fine, M. (2005). Not in our name: Reclaiming the democratic vision of small school reform. *Rethinking Schools* 19: 11–14.
Franklin, B.M. (1986). *Building the American community: The school curriculum and the search for social control*. London: Falmer Press.
Grant, G. (1988). *The world we created at Hamilton High*. Cambridge: Harvard University Press.
Green, A. and Wiborg, S. (2004). Comprehensive schooling and educational inequality: An international perspective. In M. Benn and C. Chitty (eds.), *A tribute to Caroline Benn: Education and democracy* (pp. 217–42). London: Continuum.
Kliebard, H.M. (2004). *The struggle for the American curriculum, 1893–1958* (3rd ed.). New York: Routledge.
Krug, E.A. (1964). *The shaping of the American high school, 1889–1920*. Madison: University of Wisconsin Press.
———. (1972). *The shaping of the American high school, 1920–1941*. Madison: University of Wisconsin Press.
Labaree, D.E. (1988). *The making of an American high school: The credentials market and the Central High School of Philadelphia, 1838–1939*. New Haven: Yale University Press.
McCulloch, G. (2004). *Documentary research in education, history and the social sciences*. London: Routledge.
———. (2007). *Cyril Norwood and the ideal of secondary education*. New York: Palgrave Macmillan.
McCulloch, G. and Richardson, W. (2000). *Historical research in educational settings*. Buckingham: Open University Press.

McNeil, L.M. (1986). *Contradictions of control: School structure and school knowledge.* New York: Routledge and Kegan Paul.

———. (2000). *Contradictions of school reform: Educational costs of standardized testing.* New York: Routledge.

Meier, D. (1993). *The power of their ideas: Lessons for America from a small school in Harlem.* Boston: Beacon Press.

———. (2002). *In schools we trust: Creating communities of learning in an era of testing and standardization.* Boston: Beacon Press.

OECD (2003). *Learning for tomorrow's world: First results from PISA.* Paris: OECD.

Powell, A., Farrar, E., and Cohen, D. (1985). *The shopping mall high school: Winners and losers in the educational marketplace.* Boston: Houghton & Mifflin.

Ravitch, D. (2000). *Left back: A century of failed school reform.* New York: Simon & Schuster.

Reese, W.J. (1995). *The origins of the American high school.* New Haven: Yale University Press.

Savoie, P., Bruter, A., and Frijhoff, W. (eds.) (2004). *Secondary education: institutional, cultural and social history.* Special issue of *Paedagogica Historica* 40: 1–2.

Sedlak, M.W., Wheeler, C.W., Pullin, D.C., and Cusick, P.A. (1986). *Selling students short: Classroom bargains and academic reform in the American high school.* New York: Teachers College Press.

Tanner, D. and Tanner, L. (1990). *History of the school curriculum.* New York: Macmillan.

The American Diploma Project (2005). *Ready or not: Creating a high school diploma that counts* (Executive Summary). Washington, D.C.: National Governors' Association.

Toch, T. (2003). *High schools on a human scale: How small schools can transform American education.* Boston: Beacon Press.

World Bank. (2005). *Expanding opportunities and building competencies for young people: A new agenda for secondary education.* Washington: World Bank.

Wraga, W.G. (1994). *Democracy's high school: The comprehensive high school and educational reform in the United States.* Lanham: University Press of America.

Historical Perspectives

Chapter 2

Custodialism and Career Preparation in a Comprehensive High School, 1929–1942

Sevan G. Terzian

Historians of the American high school have characterized the Great Depression and World War II years as a pivotal period when academic and vocational preparation gave way to a new custodial mission. The shrinking of the youth labor market led some government officials and educators to question the efficacy of vocational education. New federal programs for youth through the Civilian Conservation Corps and National Youth Administration presented educational alternatives for American adolescents. Faced with growing enrollments and budget shortfalls, secondary school educators found themselves on the defensive. In response, they attempted to accommodate all youth by focusing on students' immediate needs and implementing a general academic curriculum with less rigorous courses that included subjects such as descriptive biology, problems in American life, or personal service or standards classes. This custodial mission "shifted the entire thrust of secondary education away from what had been its purposes in the first quarter of the 20th century—preparation for either college or work—and replaced them with a far more nebulous purpose, keeping teenagers in school as long as possible" (Angus and Mirel, 1999, p. 98). According to these accounts, educators' attempts to keep youth out of the shrinking labor market initiated a pattern of academic decline in American secondary education. By compromising academic standards, custodialism did not prepare youth adequately for employment (or presumably, higher education) and thus weakened the economic value of the comprehensive high school (Angus and Mirel, 1993, 1999; Cohen, 1985; Krug, 1972).

Custodialism has also carried a different connotation: as preparation for democratic citizenship. The *Cardinal Principles of Secondary Education* report of 1918 marked the extracurriculum of the comprehensive high school as a vehicle for civic preparation: "the prototype of a democracy in which various groups must have a degree of self-consciousness as groups and yet be federated into a larger whole through the recognition of common interests and ideals. Life in such a school is a natural and valuable preparation for life in a democracy." (quoted in Hammack, 2004, p. 9). This quest motivated some high school educators during the Great Depression: "Not everyone advocated the custodial function simply to get unwanted youth off the labor market. Some felt a sense of responsibility for youth who in any case would be unable to find jobs, while others sincerely considered the high school the best place for all youth to be" (Krug, 1972, p. 311). By housing diverse students under one roof and allowing them to intermingle in clubs, teams, and school-wide assemblies, the comprehensive high school aimed to serve a democratizing function (Rury, 2002). The rise of general education, moreover, complemented educators' efforts to teach students how to reconcile individual and societal needs (Wraga, 1999, p. 528). This "unifying" element of the comprehensive high school—through a general academic curriculum and extracurricular activities—would cultivate youth for their roles as democratic citizens.

To explore both the academic and civic dimensions of custodialism in the American comprehensive high school, this chapter focuses on curricular, administrative, and social developments at one institution of secondary education—Ithaca High School (IHS) in Ithaca, New York—from 1929 to 1942. To what extent did custodialism emerge at the sole comprehensive high school in the community during a period of significant growth and curricular change? The answer to this question depends on whether we consult the perspectives of school officials or students, and whether we think of the comprehensive high school's mission in primarily academic or civic terms. From the viewpoints of leading school administrators in Ithaca—based on their annual district reports, meeting minutes, and bulletins—the preparatory function of IHS remained prominent during the Depression years. Although a wholesale curricular revision created a host of new choices for students and paths to graduation that did not require rigorous academic courses, the superintendent and guidance department staff still sought to prepare students for college (mainly Cornell University) or for jobs in the restricted local economy. According to some former students, however—as revealed by their comments in follow-up surveys—the high school had not prepared them effectively for college or the job market. Some students writing in the school newspaper, meanwhile, complained about the absence of civic custodialism: IHS was not preparing its students for life in a democracy. Taken together, the expressed views of school officials and students yielded a mixed review of the ability of the sole comprehensive high school in Ithaca to prepare local youth for either college or employment while cultivating democratic values and skills for all.

The Local Context: Ithaca during the Great Depression

Situated in the Finger Lakes region of New York State, Ithaca occupied the seat of Tompkins County. Ithaca had 20,708 residents in 1930, 87.2 percent of whom the U.S. Census classified as "native white," with 9.5 percent "foreign-born white" and 3.1 percent "Negro." Agriculture was a prominent part of the county's economy, and other major sources of employment in the county included iron and steel industries, various professional and semiprofessional services, and wholesale and retail trade. In 1930, 5.8 percent of the local workforce was unemployed. Cornell University's impact on local hiring, meanwhile, was negligible. It recruited its faculty globally and enrolled thousands of students from across the nation each year. A decade before World War II, the university was far from Ithaca's largest employer of local residents (U.S. Bureau of the Census, 1931, 1932). Many adolescents in Ithaca did not work, and the proportion of Ithaca's teenagers attending secondary school well exceeded state and national averages. In 1930, 98.0 percent of 14–15-year-olds, 75.8 percent of 16–17-year-olds, and 43.0 percent of 18–20-year-olds attended school (U.S. Bureau of the Census, 1932, 1933).

By 1940, Ithaca's population had dropped to 19,730. Its ethnic and racial demographics had not changed appreciably. Agriculture remained a prominent source of occupation in Tompkins County, but of the 2,212 county residents working on farms, a mere 64 lived in Ithaca. The highest occupational category in Ithaca was clerical and sales (which also employed the largest number of females), followed by professional workers, and craftsmen and foremen. The educational attainment of Ithaca's students in 1940 again well exceeded state and national averages. While 95.9 percent of 14- and 15-year-olds attended school, there was relatively little attrition by the high school years, as 87.8 percent of 16- and 17-year-olds and 47 percent of 18–20-year-olds enrolled (U.S. Department of Commerce, 1975; U.S. Bureau of the Census, 1943).

An occupational study of nearly a thousand local workers and some employers issued in 1941 by the Junior Chamber of Commerce reveals more about Ithaca's employment structure on the eve of World War II. Addressing local education boards, the survey aimed to determine job opportunities and the necessary vocational training in the high school.

Employers lamented that they devoted too much time and resources to training their employees, and some faulted the local high school for failing to impart skills including "the ability to write well formed and complete sentences," "ability to employ salesmanship in human relationships," and good "penmanship" (Junior Chamber of Commerce, 1941, p. 5B). Not all expressed criticism, however: "Many employers complimented the schools on the general education and specific training given the young people in Ithaca. The general note of the majority of the comments was optimistic" (Junior Chamber of Commerce, 1941, p. 5C). The report indicated that there was a local demand for more trained clerks, typists, and stenographers; mechanics; and salesworkers. Employers then commented on eight types of jobs and the commensurate educational training required. In their estimation, some positions for "industrial machine operators," "automobile service," and "laborers" did not warrant secondary schooling, but most occupations in "office work," "service work," and various "trades" required a high school education. Employers indicated that work in "sales" and "professions" necessitated a high school education and beyond (Junior Chamber of Commerce, 1941). From the perspective of those local employers whose views this report summarized, nearly all of the jobs in Ithaca required some amount of high school education. It is also apparent that these employers expected the high school to assume the primary responsibility of preparing youth for such jobs. To what extent, then, had the only public high school in the community—Ithaca High School—attempted to prepare its students for their future careers?

Administrators' Perspectives: Vocational and College Preparation

Founded in 1875 by the newly created board of education, Ithaca High School served local and nonresident adolescents, many of whom sought admission to the young, but increasingly popular local institution of higher education: Cornell University. Enrollments at IHS grew steadily over the next five decades, and despite the addition of some commercial and vocational courses to its burgeoning curriculum, its focus remained primarily academic, as more than half of its graduates went on to college. By the eve of the Great Depression, Ithaca High School resembled a comprehensive institution. An unprecedented high of 1,327 students enrolled in the fall of 1929, and the high school's curriculum included academic and nonacademic courses of study. The school district's superintendent of three decades, Frank David Boynton, argued that a modern public school program warranted a differentiated curriculum. Unlike some educators of his day, however, Boynton believed that parents, not school officials, would choose the future vocation and appropriate curriculum for their children:

> The child whose parents desire him prepared for college, e.g. takes English for four years, mathematics for at least two (depending on the college selected), usually two foreign languages, history and science; the pupil whose parents desire him to enter business takes the commercial group, less or no foreign language, a different selection of mathematics, etc.; and the child whose parents desire him to enter the crafts takes the bulk of his work in the industrial or household arts groups.

To assist in strengthening vocational guidance, the superintendent even invited parents to evaluate the adequacy of existing school subjects (Boynton, 1929, p. 137). In short, curricular differentiation would accommodate the varying aspirations of families, rather than the measured abilities of pupils.

Over the course of the next decade and a half, however, a change in administrative leadership, coupled with a massive rise in enrollments and weakening job market for local youth, prompted a series of developments that affected the preparatory functions of IHS. Boynton's death from an automobile accident in 1930 led to the promotion of Claude Kulp as superintendent. Like his predecessor, Kulp advocated curricular differentiation at the high school. Unlike Boynton, the new superintendent believed that school officials should determine students' probable future destinies and courses of study. This conviction informed important institutional transformations. First, booming student enrollments from 1929 to 1935 prompted the construction of new physical facilities. Second, Kulp's belief that nearly universal enrollments revealed wide variations in abilities among students fueled substantial revisions to the high school curriculum. Third, these curricular changes at IHS justified the creation and expansion of a guidance department and placement bureau, which reflected the assumption that school officials, not parents, should steer students into appropriate courses of study. Most of these developments attempted to further the preparatory functions of the comprehensive high school.

From 1929 to 1935, student enrollments at IHS ballooned from 1,247 to 1,862, an increase of nearly 50 percent, while the city's population remained relatively constant (Ithaca Public Schools, 1935). Such dramatic growth was not unique to the high school in Ithaca, as the Great Depression pushed large proportions of adolescents out of the job market (U.S. Department of Commerce, 1975). Unlike many school districts, however, Ithaca's did not endure severe budgetary problems, especially in the first half of the 1930s. The district did not attempt to raise local school taxes until late in the decade. Annual Reports and Superintendent's Bulletins suggested that conservative spending habits and the educational contributions from the State of New York prevented cuts in teacher salaries or courses (Kulp, 1931, 1932, 1933; Ithaca Public Schools, 1932). Local economic conditions were far from ideal, however, as many other workers lost their jobs or experienced pay cuts (Kulp, 1931, 1933).

These financial circumstances allowed the district to complete several projects that expanded the physical facilities of the secondary schools. In 1928, the board of education had unanimously passed a resolution to construct a junior high school facility "to meet the normal growth of our elementary and high school grades," and because "the use of the annexes and the scattering of the Junior High school years in four different buildings makes it impossible to have a Junior High school organization" (Board of Education, 1928, p. 203). It also approved the construction of a cafeteria in the high school to prevent students from leaving during the lunch hour, a facility that could be construed as "custodial" (Graebner, 2001). In 1932, the superintendent updated teachers about the construction project: "As a group, Ithaca teachers are fortunate, indeed. Salaries are being maintained and there is every reason to believe that our schedule will be continued, despite the fact that a splendid new Junior High School representing an outlay of a half-million dollars is rapidly nearing completion" (Kulp, 1932). The junior high school opened its doors in the fall of that year, which relieved overcrowding in both the elementary schools and the high school, the latter of which had housed "hundreds of seventh, eighth, and ninth graders" (Ithaca Public Schools, 1933, p. 21). As in many school systems in the United States, the new junior high school in Ithaca would play a role in testing and sorting students into different courses. With the junior high school building completed, Ithaca's superintendent initiated a wholesale reevaluation of the senior high school's curriculum to further its preparatory functions: for vocations and higher education.

Early in his tenure as superintendent, Kulp attributed the rapid rise of student enrollments at Ithaca High School to the effects of the Depression on the local economy: "Children who would be working in normal times, were in school because there was no work" (Ithaca Public Schools, 1931, p. 40). He also believed that many of these nontraditional high school students possessed intellectual limitations: "Obviously all of these pupils are not capable of completing a conventional four year high school curriculum which deals primarily with preparation for college or business" (Ithaca Public Schools, 1931, p. 44). Kulp assumed that these students did not aspire to nor were capable of going to college. For instance, he called for courses in vocational agriculture to accommodate the growing presence of students from rural parts of the school district. He also proposed that the high school establish variable graduation criteria: "Some should not be expected to conform to the traditional four year requirements, but should be given an opportunity to enroll in curricula leading to graduation in one, two, or three years" (Ithaca Public Schools, 1933, p. 33). Believing that school officials needed to learn more about the relationship between employment opportunities and the high school's programs of study, Kulp appointed a committee of teachers and administrators to determine the extent to which the existing curriculum prepared students for the demands of the local economy (Ithaca Public Schools, 1932).

Like most comprehensive high schools in the early twentieth century, the curriculum at IHS became more specialized and proliferated beyond academic subjects. By the 1932–1933 academic year, when Kulp's committee began its assessment, the high school's curriculum included nine separate courses of study, each of which scripted a student's schedule for all eight semesters. The Classical, Modern Language, Engineering, Music-Academic, and Normal Entrance courses of study prepared students for different possible majors at institutions of higher education. Their required amount of classical and foreign languages, mathematics, and science courses varied to some extent. The courses of study that did not lead to college included "Academic" (for future nurses, pharmacists, or homemakers), Commercial, and Art. Finally, the Industrial Arts major, established in 1932 and designed for those who did not plan to go to college and did not want to take commercial subjects, required students to obtain permission from the principal of the high school to gain admission. Its curriculum did not include courses beyond "general science" and "commercial arithmetic." Enrollment figures for seniors at IHS for the 1932–1933 academic year reveal that the Modern Language major enjoyed the greatest popularity with 31 percent of the senior class. The Commercial major ranked second with 23 percent of the students enrolled. Only 8 percent of students that year majored in the Classical course of study, and a mere four students (1.88 percent) enrolled in the Industrial Arts major. Of the 213 seniors at IHS that year, 102 (or 48 percent) gained admission to an institution of higher education, a figure well above the state's average, but the lowest percentage in the school's history (Bigham, 1932).

In 1933, officials at IHS abolished the nine existing courses of study and their corresponding high school diplomas. In their place, they implemented a system of core courses and electives in a fashion that resembled some of the recommendations of the *Cardinal Principles of Secondary Education* report of 1918. All students were required to take four years of English, physical education, and "guidance." The new curricular arrangement also mandated one year of general science, social studies, and American history. Mathematics courses were no longer required. The vast remainder became electives, with subjects ranging from fourth-year Latin and chemistry to commercial arithmetic and advanced metal shop. This marked a dramatic departure from the highly structured criteria for the previous nine courses of study. This rearrangement, moreover, accompanied the establishment of a guidance department that would attempt to play an increasing role in shaping students' curricular choices and vocational goals (Ithaca Public Schools, 1934).

The curricular reorganization at IHS also seems to have facilitated the grouping of students by ability, sometimes within the same classroom. As early as February 1930, the board of education had indicated its interest in implementing achievement tests in the high school "to correlate the ability of pupils with the work which is to be presented to them" (Board of Education, 1930, p. 292). In his first bulletin as superintendent later that year, Kulp instructed teachers and staff to "avoid all reference to the testing and grouping program employed in the Ithaca Public Schools" (Kulp, 1930, p. 1). By 1933, English teachers at the high school assigned work to students "at one of three different levels according to their ability" (Ithaca Public Schools, 1933, p. 62). This practice continued throughout the decade, and school officials implemented examinations to facilitate "the grouping of students of low ability" and customizing instruction accordingly (Ithaca Public Schools, 1940, p. 11; Ithaca Public Schools, 1938, p. 52). By 1941, students at IHS were sorted into different English and social studies classes—two of the few remaining universal requirements in the curriculum (Ithaca Public Schools, 1942). At the junior high school, meanwhile, entering students took mathematics and intelligence examinations and then were assigned a "decile ranking" to establish their "homogeneous grouping," a precedent that effectively determined a student's curricular emphasis once in high school (Bartholomew, 1942, pp. 15–16). In 1936, moreover, the high school had begun to award certificates for graduation in lieu of diplomas to those students "who have not been able to meet the usual Regents requirements, but who have completed 16 units of work in courses designed for groups of different ability levels" (Ithaca Public Schools, 1936, p. 24; Board of Education, 1935, p. 175). These developments suggest that school officials intended to apply differentiation and ability grouping to minimize dropout rates.

Did these curricular reforms further the high school's preparatory functions or mark the presence of a custodial mission? On the one hand, faced with a weakening local economy for youth and rising enrollments in the first half of the 1930s, Kulp and school officials in Ithaca may have wanted to find ways to prevent students from dropping out. A local provision required all youth under 17 to remain in school unless gainfully employed. The State of New York also required public schools to enroll unemployed adolescents in special school programs. Furthermore, dozens of IHS students who had completed the twelfth grade were enrolling as "postgraduate" students, because they were unable to find a job or gain admission to an institution of higher education. Finally, the proliferation of electives in the new curricular plan at IHS could allow some students to complete their studies without taking a single mathematics, advanced science, or foreign language course. By introducing nonacademic subjects, implementing flexible (and less academically rigorous) requirements for graduation, and some ability grouping, this comprehensive high school began to exemplify some custodial features during the Great Depression.

On the other hand, high school officials in Ithaca did not abandon their quest to prepare all of their students either for college or work. For instance, in requiring all students to take four years of English, they eliminated

Business English from the curriculum and no longer allowed students to use Business Writing for graduation credit. In addition, the expanded presence and role of the guidance department by the mid- to late 1930s, and the establishment of a placement bureau, indicated that teachers and administrators at IHS wanted to learn from former students whether they had succeeded in preparing them for their vocations or college careers. In these ways, custodialism did not appear to have been the intent of school officials. To what extent did students at IHS agree? Two follow-up surveys issued by the guidance department of IHS, as well as issues of the student newspaper, illustrate the sentiments and insights of the recipients of these curricular changes.

Students' Perspectives: Inadequate Preparation for Careers and Citizenship

In 1936 and 1942, the guidance department of IHS issued surveys to its former students from the Classes of 1935 and 1941 to determine which high school courses had been most "useful" for their current employment or studies. Guidance department staff sought to minimize dropout rates and believed that students often withdrew because of a "lack of adjustment in school or the limited offering of our school" (Meyn, 1936, p. 28). The response rates were impressive. Over 90 percent of the graduates and 86 percent of nongraduates responded to the 1936 survey, while 83 percent overall responded to the 1942 survey. By asking former students which courses they had found most useful, school officials sought to alter the high school curriculum to prepare current students more effectively for either college or the workplace (Meyn, 1936, p. 7; Ithaca Public Schools Guidance Service, 1942).

The 1936 follow-up survey solicited responses from 312 graduates from the Class of 1935 as well as 181 former students who had withdrawn before graduation. Graduates most frequently listed academic courses (such as English, mathematics, and science) as the most useful to their present studies or employment. Members of the Class of 1935 had begun their high school careers under the old system of the separate courses of study and their corresponding diplomas. Fifty-five percent of them had prepared for college. The Commercial major followed in popularity (32.4 percent of graduates), while 24.3 percent of 1935 graduates had majored in the Modern Language course. Less than 3 percent had majored in the Vocational course of study, while less than 1 percent had majored in the Industrial Arts. These responses led the guidance committee to conclude that "the traditional subjects still are considered most important, making suggestions as to revision or curtailment of such subjects rather non-practical"

(Meyn, 1936, p. 15). It also acknowledged that it was difficult to generalize how former students earning wages obtained their current employment and that "a job, regardless of its kind or possibilities, is acceptable to a high school graduate these days." Nonetheless, it maintained, "vocational or prevocational training seems to have a definite bearing on placement" and called for enlarged functions of the placement bureau (Meyn, 1936, p. 16). In short, the guidance department sought to extend its preparatory mission.

Two hundred and ninety of the 348 members of Class of 1941 responded to the second follow-up survey. Of the 164 respondents who had pursued an "academic" sequence for college preparation, 96 indicated that they were enrolled in school full time (the majority at Cornell University), while 34 were working for wages and 11 were unemployed. Of the 92 respondents who had pursued a commercial sequence in high school, 58 indicated that they were employed for wages, while 8 were unemployed. The 3 most prevalent sources of employment for the 111 graduates of the class of 1941 were clerical office work, factory and shop work, and work in stores (Ithaca Public Schools Guidance Service, 1942).

Respondents to the 1942 survey typically indicated that the guidance department at IHS had not played a salient role in shaping their educational and vocational plans. When asked who had helped them obtain their first job, for example, 40 percent indicated that it was from their own effort and 19 percent attributed it to a friend. Only 12 percent cited teachers or school counselors. When asked to evaluate the effectiveness of their vocational training at IHS for their present employment, a mere 18 percent indicated that it had helped directly, while 38 percent wrote that it had helped in a "general way only" and 34 percent wrote that it had "provided some training." Ten percent said it had not prepared them at all for their current work. The primary value of guidance counseling at IHS, according to 57 percent of the respondents from the Class of 1941, had been in helping them select their high school courses. Only 10 percent found school counseling helpful in applying for a job, and only 9 percent believed that it had helped them prepare for their future studies. Indeed, 39 percent of respondents had wished that the high school had given them more specific job training, and 36 percent wanted to have received more help in planning their future education (Ithaca Public Schools Guidance Service, 1942).

Perhaps the greatest indictment of the guidance department, and overall preparatory function of IHS, is indicated in the respondents' reliance on their parents in determining their educational or vocational ambitions. Even guidance officials acknowledged this trend: "Parents loomed up as the most important advisers in assisting students in making their educational plans. To fail to make the parents an important part of this process would be to omit a key factor. No other adviser apparently has the same influence." Half of the respondents said that parents had helped them the most in determining their educational plans; only 5 percent cited their school counselors. Moreover, while 41 percent of respondents identified their parents as most influential in determining their vocational goals, less than 1 percent listed school counselors as most influential. The relative insignificance of guidance counseling suggests that in the eyes of former students from the Class of 1941, IHS had not been particularly successful in preparing them for their educational or vocational goals. In addition, despite Superintendent Kulp's and the board of education's efforts to augment the high school's role in steering student coursetaking and career aspirations, families retained a primary role (Ithaca Public Schools Guidance Service, 1942).

Most respondents to both follow-up surveys were either silent or critical of the high school's ability to prepare them for their current endeavors. Examples of praise were far less common. The responses of former students also varied considerably, particularly between those in college and those not in college. Nearly all of the graduates from 1935 and 1941 attending college criticized IHS for not having prepared them adequately for the rigors of higher education. Some expressed a disdain for their high school English courses. A former student studying biochemistry and nutrition at Cornell University, for instance, complained that "the English course is of no good in preparation for college. A *thorough* background in *composition, grammar* and *spelling* is one of the *greatest* improvements that could be done" (Meyn, 1936). One respondent to the 1942 survey accented the gap between high school and college: "Cornell and high school don't agree very well. I think English classes for future college students in which they would receive vital basic training would be of great value" (Ithaca Public Schools Guidance Service, 1942, p. 2). Another reported that his IHS peers at Cornell were encountering great problems: "In fact we go to college so poorly prepared in English that the instructors claim they can pick out the Ithaca students in the first few days by what they know" (Ithaca Public Schools Guidance Service, 1942, p. 9). For such respondents, high school English courses that better prepared them for the rigors of college constituted a patently "useful" suggestion.

Former IHS students attending college also expressed concerns about their lack of self-discipline and initiative, and they blamed the high school for their difficulties. One respondent wrote in 1936 that "high school fails to teach students how to study as they must when they get to college... [and]... could do much more to breach the gap between high school and college" (Meyn, 1936). Numerous former students complained that they suffered from poor study habits and urged the high school to address this matter directly in its curriculum. Accordingly, some offered specific suggestions about what sorts of courses and policies should be implemented. One recommended that a specific orientation course be introduced and "taught by a recent graduate of a large institution like Cornell... which would acquaint innocent lambs with the facts about a possible slaughter house if they don't learn how to study, be independent, take notes and above all *organize* everything" (Meyn, 1936). A 1935 graduate from IHS preparing for medical school at Dartmouth College echoed these sentiments: "In High School one is guided, forced to do his work and punished if he doesn't do it . . . Why not make Ithaca High an exception and really prepare the students for college" (Meyn, 1936)? In these ways, some former students lamented that they had not received more explicit college preparation in both their coursework and their teachers' academic expectations at IHS. They expressed resentment that they had not been "taught how to study efficiently." It is somewhat ironic that while these former students wished they possessed greater independence and self-discipline for the rigors of college work, they advocated the addition of high school courses geared toward that end. Self-discipline and study habits were not to be cultivated through one's own efforts, but were seen to be part of the preparatory mission of the high school.

Like those in college, a number of graduates who had joined the workforce or were searching for employment wished that their former high school had forged closer connections between its curriculum and their current endeavors. These respondents expected the high school to have prepared them to succeed in a limited job market. One former student, for example, did not think that the high school had succeeded in its vocational preparation: "It would be a great advantage to pupils who cannot attend college if the Ithaca High could teach them a trade or some specialized vocation. It is very difficult for the present high school graduates to obtain worthwhile positions as they are not suited for any" (Meyn, 1936). Some graduates wanted the high school to improve its methods of career counseling: "It is a sad experience to find yourself a graduate with no definite vocation in mind . . . Cannot some system be arranged whereby students may be urged to formulate some idea of what their future vocation is to be?" (Meyn, 1936). Others wished that the high school had better acquainted students with future career possibilities whether through informational lectures or field trips to various workplaces (Ithaca Public Schools Guidance Service, 1942; Meyn, 1936). Many respondents to both surveys suggested that the high school increase its specialized training in commercial and industrial courses. Perhaps one IHS graduate in the workforce encapsulated the primary concern of his cohorts, when he proposed that the high school would be of greater value "if the courses were made to seem real" (Ithaca Public Schools Guidance Service, 1942, p. 5). From the perspective of some former students either employed or looking for work, the high school had not fulfilled its preparatory mission. While "usefulness" meant specific vocational training, career placement, and even familiarity with working environments, these respondents echoed their peers attending college in calling for IHS to bridge the gap between secondary studies and life beyond. As a result, they called for the expansion of the placement bureau and further curricular specialization: to strengthen the high school's preparatory mission.

Even as the wartime economy began to reduce unemployment by the early 1940s, some former students at IHS wished that the guidance department had been of greater assistance. One complained about the impersonal treatment he received from the guidance staff: "The counselors now don't seem interested enough in the pupils. It is just more or less a title and it is almost impossible to see a guidance teacher when you want to and when you do they don't give a darn about your problems, just give some advice to get rid of you" (Ithaca Public Schools Guidance Service, 1942, p. 7). Another respondent lamented: "It seems to me that the only counseling I received was to make sure that i [sic] had enough units to graduate" (Ithaca Public Schools Guidance Service, 1942, p. 15). Another former student wished he had been given "more personal guidance; more individual conferences with advisors and suggestions and explanations of them. To me they were of absolutely no value throughout my entire schooling" (Ithaca Public Schools Guidance Service, 1942, p. 13). Yet another expressed frustration that "my counsel has not helped in anything at all. Have asked for help and they just say come back again and its [sic] the same old story—finally giving up" (Ithaca Public Schools Guidance Service, 1942, p. 3). These overt criticisms about the guidance department at IHS, coupled with the various frustrations of former students either in college or the job market, exemplify the enormous challenges facing a single comprehensive high school that sought to prepare all of the adolescents in the community for the local labor market or the rigors of higher education. In the eyes of some former students, IHS had failed to provide the attention and services they deemed necessary to have prepared them to succeed in their current endeavors and to meet their career aspirations.

Part of the challenge facing the counseling and teaching staffs at this comprehensive high school in an era of economic depression and war was meeting the diverse ambitions and needs of over a thousand students. Because of the curricular plan of constants and electives spurred by Superintendent Kulp and initiated in the fall of 1933, the newly formed guidance department would seemingly play a prominent role in steering students into appropriate courses in accordance with their future aspirations: whether for admission to Cornell University or other institutions of higher education, or entry to a trade or gainful employment in the local economy. The two guidance department follow-up surveys from 1936 and 1942 in particular illustrate the enormous range of opinions regarding which courses students deemed "useful." While some cited typing and shorthand as most useful, others prized physics, history, and music.

In addition, the guidance department's lofty claims about its services appeared in places such as the students' high school annual—and may have raised students' expectations about its efficacy. A page in the 1936 yearbook, for example, proclaimed that the guidance department would advise students on which courses to take for admission to Cornell; steer students into specific high school courses; teach them how to study more effectively and improve their grades; help students choose appropriate extracurricular activities; and even explain to a student why he or she was "not more popular." In sum, the guidance department claimed that it could assist all students in all facets of their lives and answer the question of "What shall I be?" (Ithaca High School, 1936, p. 87). Given this heightened rhetoric, respondents to both follow-up surveys expected the high school to have equipped them with the knowledge and skills either to succeed in college or to secure gainful employment. It is not altogether surprising, moreover, that some former students—whether in college or in the workforce—expressed disappointment that IHS had not provided sufficient preparation for their careers. While it was not the intent of school officials, some former students seemed to indicate that the high school had assumed a custodial function in an academic sense.

Did students at IHS also see civic custodialism in evidence? Did they think that the social life of the high school had prepared them for democratic citizenship? Evidence from the school newspaper, *The Tattler*, suggests that the extracurriculum at IHS did not typically foster the social intermingling of an increasingly diverse student body. In editorials and letters to the editor, students complained about the absence of or minimal support for student government; an elite and poorly supported array of student clubs; and social divisions among the student body. One staff editorial, for instance, lamented the failure of a student government in years past and urged its implementation: "It aids us in becoming better citizens. From it, we learn independence, promptness, and leadership" ("Student Government," March 24, 1932, p. 2). A student letter to the editor three years later revealed that no progress had been made in the matter and expressed concern that the high school was not preparing students for citizenship: "We will not always have things taken care of for us, so let us start now" ("Student Government," January 16, 1935).

In addition to student government, architects of the comprehensive high school in the early to mid-twentieth century promoted school clubs as a way to foster social intermingling among students of different backgrounds. Despite the presence of dozens of social, athletic, and academic organizations at IHS, however, they were not always well supported or seen as democratic. In a letter to *The Tattler*, IHS teacher and guidance counselor, E.A. LaFortune, criticized their elitist tendencies: "Some clubs owe their continued existence to a false sense of prestige which is supposed to ensue to the pupil upon becoming affiliated with the club" ("School Clubs," April 17, 1935, p. 2). LaFortune complained that the policy of nominating members fostered discrimination and recommended instead that measures of academic performance determine a student's eligibility to join. This exclusiveness may have contributed to an apparent lack of student interest in clubs by the late 1930s. Although clubs held "a very important position in school life" and had the potential to "work for the betterment of acquaintanceship between the students," editors at *The Tattler* complained that many clubs "have remained practically inactive"

("Clubs," May 28, 1936, p. 2). A subsequent staff editorial revealed that poor student attendance at club meetings persisted, and that membership still required a nomination or invitation ("Too Many Extra-curricular Activities!" October 31, 1940).

The perspectives of some former and current students at IHS also suggest that extracurricular activities did not mitigate social and class divisions within the school. A student's letter to the editor in *The Tattler* alluded to the residential division in Ithaca between the "Hill Group" (middle-class families largely affiliated with Cornell University) and the "Valley Group" (primarily working-class and immigrant families working in local industries). This student drew a parallel to the student body at IHS: "Each floor in our school [is] in a different family or group. These groups have Civil Wars between themselves" ("Sectionalism," December 5, 1935, p. 2). Several respondents to the 1942 follow-up guidance survey, moreover, argued that such divisions restricted opportunities. One contended that "even the teachers and counselors show favoritism toward a certain class" (Ithaca Public Schools Guidance Service, 1942, p. 3). Another complained that the administration lacked "enough interest . . . for the average pupils" (Ithaca Public Schools Guidance Service, 1942, p. 12). According to these current and former students, social schisms in this comprehensive high school reflected class divisions in the community. Student government was absent or invisible, and clubs were elitist and poorly supported. Instances of student praise of these organizations in the student newspaper were virtually nonexistent. From these examples, therefore, it appears that the extracurricular dimension of this comprehensive high school did not foster the type of civic preparation promoted by some Depression-era educators.

Conclusions

In tracing the history of the American high school, historians David Angus and Jeffrey Mirel (1999) point to the Great Depression and World War II as a turning point: "High schools shifted from institutions concerned primarily with academic and vocational preparation to institutions largely responsible for custodial care" (p. 57). This statement does not completely characterize the situation at Ithaca High School. On the one hand, Superintendent Claude Kulp and other school officials did not attempt to alter the school's mission to prepare students for jobs or higher education. The establishment of a guidance department and placement bureau indicates that the preparatory function of the comprehensive high school in Ithaca persisted through the 1930s and early 1940s. The guidance department's quest to gauge its effectiveness (and the school's curriculum) in preparing former students for college or the workplace furthers the point that school officials did not intend to shift the school's mission to academic custodialism. As far as civic custodialism was concerned, moreover, some students writing in the school newspaper did not believe that the extracurriculum at IHS fostered the sort of student involvement that would prepare them for democratic citizenship. On the other hand, the curricular reorganization of 1933 allowed some students to choose less academically rigorous paths to graduation—one of the hallmarks of academic custodialism. Furthermore, the criticisms of many of the former students to the follow-up surveys of 1936 and 1942 suggest that, whatever the intentions of school officials, the high school was not succeeding particularly well in preparing youth to succeed in their future endeavors. From their perspectives, inadequate preparation amounted to little more than academic custodialism.

This historical account of the multiple manifestations of custodialism—both academic and civic—can shed light on current criticisms about the comprehensive high school in the United States. As Floyd Hammack (2004) has suggested, popular perceptions of this long-standing model of secondary education have largely been negative: "that the schools are too large and try to do too many things, none of them very well" (p. 129). According to John Rury (2002), moreover, much of the scrutiny about secondary schooling in the latter half of the twentieth century has neglected the democratizing function of the comprehensive high school: the housing of diverse students under one roof and their intermingling through clubs, teams, and school-wide events. In the nation's current political and cultural climate, secondary schools are typically not evaluated on the basis of how well they cultivate democratic citizens. Yet one of the most distinctive features of American secondary schooling has been its pageantry and central place in the community. Perhaps, then, there is and has been a valuable role for custodialism in the comprehensive high school: not as a general academic track or life adjustment curriculum, but as part of the social and intellectual cultivation of adolescents for citizenship.

Finally, this case history reveals that the multiple perspectives of educators, as well as students and their parents, must be considered in establishing the mission and assessing the relative success of any school. At IHS during times of economic depression and the onset of war, school officials attempted to augment their own influence, but parents retained an important role in shaping students' curricular decisions and career aspirations. Yet contemporary policy makers' pronouncements about high school reform rarely consider the voices of parents and students. Therefore, those who target the comprehensive high school as an educational institution ripe for redefinition must consider the perspectives and needs of the very constituents whom it serves: students, families, and community organizations. If the comprehensive high school is to survive and even flourish in the years ahead, it must do so in conjunction with, and not isolated from, these other long-standing educational agencies.

References

Angus, D.L. and Mirel, J.E. (1993). Equality, curriculum, and the decline of the academic ideal: Detroit, 1930–68. *History of Education Quarterly* 33: 177–207.

———. (1999). *The failed promise of the American high school, 1890–1995.* New York. Teachers College Press.

Bartholomew, B.M. (1942). A report of the work and activities of the first ten years in the Frank David Boynton Junior High School, 1932–1942. Ithaca Public School Archives. Ithaca, New York.

Bigham, H.R. (1932). *Hand book of the Ithaca High School, 1932–1933. Ithaca, New York.* Ithaca. Atkinson Press of Ithaca.

Board of Education. (1928). Special meeting of the board of education, 1 August 1928. In Ithaca Records, May 1925–February 1933. Ithaca Public School Archives. Ithaca, New York.

———. (1930). Board of Education Meeting Minutes, 4 February 1930. In Ithaca Records, May 1925–February 1933. Ithaca Public School Archives. Ithaca, New York.

———. (1935). Board of Education Meeting Minutes, 10 September 1935. In Ithaca Records, March 1933–March 1939.

Boynton, F.D. (1929). A modern public school program. *Ithaca Public Schools Our Point of View* XI (9). Ithaca. Ithaca Public Schools Press.

"Clubs." (May 28, 1936). *The Tattler*, vol. 7, no. 15.

Cohen, D.K. (1985). Origins. In A.G. Powell, E. Farrar, and D.K. Cohen (eds.), *The shopping mall high school: Winners and losers in the educational marketplace* (pp. 233–308) (Boston: Houghton Mifflin).

"Good Citizenship." (April 17, 1935). *The Tattler*, vol. 6, no. 12.

Graebner, W. (2001). The sexual politics of the school lunchroom, 1890–1930. *Prospects* 26: 337–60.

Hammack, F.M. (2004). Does the comprehensive high school have a future? In F.M. Hammack (ed.), *The comprehensive high school today* (pp. 129–42) (New York: Teachers College Press).

Ithaca High School. (1936). *1936 Ithaca High School Annual.*

Ithaca Public Schools. (1931). *Ithaca Public Schools Annual Report, 1930–1931.* Ithaca. Ithaca Public Schools Press.

———. (1932). *Ithaca Public Schools Annual Report, 1931–1932.* Ithaca. Ithaca Public Schools Press.

———. (1933). *Ithaca Public Schools Annual Report, 1932–33.* Ithaca. Ithaca Public Schools Press.

———. (1934). *Ithaca Public Schools Annual Report, 1933–34.* Ithaca. Ithaca Public Schools Press.

———. (1935). *Ithaca Public Schools Annual Report, 1934–35.* Ithaca. Ithaca Public Schools Press.

———. (1936). *Ithaca Public Schools Annual Report, 1935–1936.* Ithaca. Ithaca Public Schools Press.

———. (1938). *Ithaca Public Schools Annual Report, 1936–37.* Ithaca. Ithaca Public Schools Press.

Ithaca Public Schools. (1940). *School Notes for 1939–1940*. Ithaca. Ithaca Public Schools Press.

———. (1942). *Ithaca Public Schools Our Point of View, 1941–42*. Ithaca. Ithaca Public Schools Press.

Ithaca Public Schools Guidance Service. (1942). The class of '41 speaks: A follow-up study of the class of 1941 of Ithaca High School. Ithaca High School Archives. Ithaca, New York.

Junior Chamber of Commerce. (1941). An occupational study of Ithaca, New York April 1941. Ithaca Public School Archives. Ithaca, New York.

Kulp, C.L. (1930). Bulletin No. 1. Ithaca High School Archives. Ithaca, New York.

———. (1931). Bulletin No. 19. Ithaca High School Archives. Ithaca, New York.

———. (1931). Bulletin No. 23. Ithaca High School Archives. Ithaca, New York.

———. (1932). Bulletin No. 32. Ithaca High School Archives. Ithaca, New York.

———. (1933). Bulletin No. 7. Ithaca High School Archives. Ithaca, New York.

Krug, E. (1972). *The shaping of the American high school volume 2, 1920–1941*. Madison. The University of Wisconsin Press.

Meyn, A.W. (1936). Department of guidance follow-up survey 1936. Ithaca High School Archives. Ithaca, New York.

Rury, J.L. (2002). Democracy's high school? Social change and American education in the post-Conant era. *American Educational Research Journal* 39: 307–36.

"School Clubs." (April 17, 1935). *The Tattler*, vol. 6, no. 12.

"Sectionalism." (December 5, 1935). *The Tattler*, vol. 7, no. 5.

"Student Government." (March 24, 1932). *The Tattler*, vol. 3, no. 7.

"Student Government." (January 16, 1935). *The Tattler*, vol. 6, no. 7.

"Too Many Extra-Curricular Activities!" (October 31, 1940). *The Tattler*, vol. 12, no. 2.

U.S. Bureau of the Census. (1931). *Fifteenth census of the United States: Unemployment Vol. 1*. Washington, D.C. United States Government Printing Office.

———. (1932). *Fifteenth census of the United States: 1930 population volume III, Part 2*. Washington, D.C. United States Government Printing Office.

———. (1933). *Fifteenth census of the United States: 1930 population volume IV occupations, by states*. Washington, D.C. United States Government Printing Office.

———. (1943). *Sixteenth census of the United States: 1940 population volume II, Part 5*. Washington, D.C. United States Government Printing Office.

U.S. Department of Commerce. (1975). *Historical statistics of the United States: Colonial times to 1970, Part 1*. Washington, D.C. U.S. Department of Commerce.

Wraga. W.G. (1999). The progressive vision of general education and the American common school ideal: Implications for curriculum policy, practice, and theory. *Journal of Curriculum Studies* 31 (5): 523–44.

Chapter 3

Education for All American Youth (1944): A Failed Attempt to Extend the Comprehensive High School

Wayne J. Urban

This essay[1] analyzes the preparation, content, and dissemination of a publication, during World War II, of the Educational Policies Commission (EPC) of the National Education Association (NEA) and the American Association of School Administrators (AASA). It begins with a brief discussion of the comprehensive high school and then another brief discussion of the EPC and the place of *Education for All American Youth* in its larger program (Educational Policies Commission, 1944). After that, an extensive analysis of *Education for All American Youth* occurs. This analysis will show that *Education for All American Youth* attempted to extend the reach of the comprehensive approach by expanding the American high school into the first two years of post-high school education, or as called in the report, the thirteenth and fourteenth grades. This was to be accomplished by drastically expanding the curriculum of all of secondary education, including these two new grades, and by taking control of the thirteenth and fourteenth grades away from institutions of higher education and putting it into the hands of the professional educators who managed America's elementary and secondary schools.

The essay then shows that this bold attempt was short-circuited, however, by two strands of opposition. The first type of opposition came from social liberals and radicals who believed that the EPC position that real social reform could come using the schools alone was impracticable. The second set of opponents was made up of traditional academics who taught in high schools and in colleges and were opposed to the EPC position that academic content needed to be replaced, or at the least greatly supplemented by, vocational and other nonacademic studies. The second strand of criticism reached a peak with the publication of a highly successful report on secondary education sponsored by the Harvard University faculty that stole the thunder from the EPC report. The reasons for the success of the Harvard effort and the failure of *Education for All American Youth* will be explored at length, as well as the consequences of these results for the subsequent development of the comprehensive American high school. Before all of this discussion, however, a brief look at the comprehensive high school and the EPC is in order.

The Comprehensive High School and the Educational Policies Commission (EPC)

For the leaders of America's schools, the broadening of the American high school into an institution that embraced, or at least attempted to embrace, all high school age youth, began in the post–World War I era. The publication of The Cardinal Principles of Secondary Education in 1918 added objectives of social, personal, and occupational amelioration to the objective of mastery of academic subjects that characterized the traditional American high school (Urban and Wagoner, 2004). While the leaders of American education, the school superintendents, state educational officials, and normal school administrators who also dominated the NEA were firmly committed to an expanded, or what we later have come to call, a comprehensive, high school (Urban, 2000), other educational constituencies, including some high school principals, many high school teachers, especially those of the traditional academic subjects, and, even more, high school parents, were much less firmly in the comprehensive camp. In fact, it seems fair to say that the comprehensive high school, in spite of its firm institutionalization within the leadership circles in public education, has never completely carried the day in the United States. Private high schools, some public high schools with roots extending back to the nineteenth century, and many prestigious suburban high schools that acknowledge the virtues of the extended reach and diverse curriculum of the comprehensive high school, have protected academic studies, especially for their "best" students.

In this effort to protect the academic approach, usually for a "chosen" minority, college and university professors and administrators have also played a major role in curricular discussions of high schools and colleges

(Hutchins, 1936). While this protection has most often been offered in prescriptions for undergraduate higher education, it also has been a minor, but relatively constant, theme in the discourse on the American high school. For academic analysts of the American high school such as Harvard University President James Bryant Conant (1959), academic studies were often touted as especially appropriate for intellectually "gifted" students. Conant stated this prescription in a 1959 volume devoted to touting the virtues of the large, comprehensive high school, at the same time that he limited its reach by largely protecting advanced academic studies for the gifted. Conant had learned of the virtues of the comprehensive high school during several terms of service on the Educational Policies Commission. He was a member of the EPC when it published *Education for All American Youth* (Educational Policies Commission, 1944).

Founded in December of 1935, the EPC was a blue-ribbon panel of American educational leaders, cosponsored by the NEA and its most powerful affiliate, the AASA. The EPC spent its first five years outlining a social program to help ameliorate the conditions of the Great Depression through reform of American democracy, a reform that was to be led by the American public schools.[2] A significant reform agenda was outlined in the first EPC publication, written by the noted historian Charles A. Beard (Educational Policies Commission, 1937). At the end of this period, however, World War II loomed prominently on the American horizon and the EPC took the war into account in revising its own activities. In 1940, the EPC with funding from the General Education Board produced a civic education study that attempted to put the social emphasis of its first five years into a more school-oriented context (Educational Policies Commission, 1940). One year later, with the war increasingly likely to involve the United State directly, the EPC published a report drastically contrasting schooling, especially the political content of schooling, in the United States of America with education in Germany, Russia, and Japan (Educational Policies Commission, 1941). This report continued the recently emphasized, school-oriented focus for the EPC, though its agenda was more devoted to winning political support for the war from teachers and students than it was to school reform itself.

When the United States formally declared war in December of 1941, the EPC, which had a heavy k-12 focus but also always had at least a few college presidents as members, moved swiftly to contend with war's most direct consequences for the schools and the colleges of the nation. One month after Pearl Harbor, an EPC meeting considered the issue of a war policy for the schools. Specifically, the EPC attempted to head off universal military service and a draft, instead advocating a "selective service" policy that would allow some teachers to be deferred, along with allowing the best students to finish high school and/or college before entering military service (Proceedings, Educational Policies Commission, January 10, 1942; hereafter, Proceedings, EPC). At a meeting one month later, in February of

1942, the EPC entered into a discussion of secondary education that would last for the next two years. The rest of this chapter is devoted to that discussion, and the document that it produced, *Education for All American Youth* (Educational Policies Commission, 1944).

Education for All American Youth Genesis

The leading architect of *Education for All American Youth* was William G. Carr, secretary of the EPC and director of the Research Division of the National Education Association. Though Carr technically was not a formal member of the EPC, he was its secretary, or chief staff officer, and thereby wielded enormous influence over that body. He spoke infrequently at EPC meetings, but when he spoke he usually had something important to say and he expected to be heard by the members. At an EPC meeting in Berkeley, California in February of 1942, Carr described the impending report on American youth as one that would outline a major reorganization of secondary education. Such a major reorganization was not to be advocated lightly. In its effort at reorganization the EPC sought cooperation from groups such as the Youth Commission of the American Council on Education and sought financial support through a grant of $50,000 from the General Education Board, to supplement the regular support received from the NEA and the AASA. The central problem for the EPC to address in this report was the unemployment rate, particularly of young men, and the maladjustment that this unemployment indicated was prevalent in this age cohort (Proceedings EPC, February 19–20, 1942). Preparation of youth for employment had been a major, if not the major, objective of advocates of the comprehensive high school since it was proposed initially in 1918.

There was one especially important issue for the EPC that predated the war in its consideration of this problem of unemployment. The EPC had been very critical of the National Youth Administration (NYA) and the Civilian Conservation Corps (CCC), both created in the 1930s by President Franklin D. Roosevelt as New Deal agencies of the federal government to deal with youth unemployment. The onset of the war provided the EPC with an opportunity to reclaim the whole employment issue for the public schools, and to thwart the efforts of the NYA and CCC. With war preoccupying the fiscal efforts of the federal government, a well thought-out plan for employment education by the public schools would win the day for the schools against its rival government agencies. Thus, from the beginning, war provided a context within which the EPC could pursue its ends: an opportunity to continue its opposition to NYA and CCC to the point that these agencies would be abolished.

At an EPC meeting in June of 1942, one in which the group met jointly with representatives of the high school principals and the American Association of Junior Colleges, William Carr indicated that the new EPC report would deal especially with the 17–20 age group, particularly their occupational adjustment, and their leisure, health, and civic attitudes in relation to that larger objective. Another EPC official took advantage of the war context to rhetorically energize the group. He said that the new report would provide a program through which the schools would serve as a "Panzer division" for American youth. This meant that the schools would provide a whole variety of services for those youth, in addition to normal educational activities (Proceedings EPC, June 3, 1942, p. 188; Proceedings EPC, June 5, 1942, pp. 49–53). This statement reinforced an over 20-year-old priority of the American educational establishment, defense of curricular expansion in the high school and a concomitant relativization, if not denigration, of the academic subjects taught there traditionally.

During a March 1944 discussion of a draft of *Education for All American Youth*, four months prior to its publication, tension between the expansive and the traditional approaches to the high school surfaced within the ranks of the EPC itself. One member of the group raised a question about the report's lack of provision for formal instruction in English in the tenth grade. This precipitated a long discussion of the place of the traditional curriculum in the high school, with advocates of the approach outlined in the report arguing that the traditional subjects such as English and history, which were not to be offered separately, would be provided for in a "common learnings" experience in the new course of study. Common learnings was a new course that appeared in the new curriculum advocated in the report in several of the high school years, including a three-period-a-day experience in the tenth grade. In further discussion, it became clear that common learnings was to be taught by an expert in adolescent development and not necessarily in one of the traditional subjects. Thus tenth graders would be left without any full experience in *traditional* subject matter content, instead receiving an interdisciplinary, common learnings course that had no formal academic orientation and stressed, instead, more on vocational or exploratory experiences.

One EPC member stated that while something like common learnings was appropriate for many students, especially those not now served by the high school, some students would want and need the traditional subjects. The eventual compromise reached was a reduction in common learnings from three periods per day to two in the tenth grade, and addition of a separate period of science instruction. Also added were stipulations that common learnings was to be organized experimentally and that separate subjects such as English and history were to be taught within common learnings only in their general aspects (Proceedings EPC, March 11, 1944, pp. 261–309; Hampel, 1986, pp. 37–41).

Clearly, the EPC was waging a battle against the traditional, disciplinary high school curriculum. One member identified Robert Hutchins (1936), whose ringing defense of traditional academic study had been well received in academic and in lay circles, as an intellectual reactionary who cared not at all for all American youth. The EPC, on the other hand, was committed to educating all youth but intent on a radical revision of the high school curriculum to accomplish this objective (Proceedings, EPC, March 11, 1944, pp. 404–06). Pursuant to that revision, the EPC also noted that individual differences within the high school students needed to be acknowledged programmatically, thereby ensuring that the traditional academically oriented subjects would be preserved for the best students, if not for all students. In a nod to opponents of such grouping as too rigid, the EPC agreed that individual differences were not to be frozen in a policy of ability grouping (Proceedings, EPC, March 11, 1944, p. 160). As was frequently the case in discussions of this nature, the ways to avoid freezing the students in groups once they were placed there were not discussed.

These discussions revealed curricular expansiveness as a comprehensive, core of the EPC's approach to the high school. As just seen, however, even within the EPC itself a minor, but important, academically oriented criticism of that approach surfaced. As we will see shortly, that criticism would be echoed, but much more stridently, after the publication of *Education for All American Youth*. Before addressing that criticism, both within the EPC and outside of it, however, a look at the entire body of the document is appropriate.

Education for All American Youth Contents

The published version of *Education for All American Youth* was more than 400 pages long. In those pages, the EPC outlined its provision for secondary education in a postwar America, thereby simultaneously acknowledging and relativizing the wartime context in which the book had been prepared. While the report never mentioned the existence or desirability of what we have come to call a "comprehensive" high school, it did point to a significant part of the comprehensive agenda in its title: the education proposed in its pages was to be for *all* American youth, not just for those oriented to or interested in traditional academic study. In explicating the words of the title, the report noted: "When we write confidently and inclusively about education for *all* American youth, we mean just that. We mean that all youth, with their human similarities and equally human differences, shall have educational services and opportunities suited to their personal needs and sufficient for the successful operation of a free and democratic society" (Educational Policies Commission, 1944, p. 17).

The report began with a dystopian look at a future educational system in which local schools drifted along, mainly because their funding was strangled by lack of support from the federal government. In this dismal future, rather than support the schools, the federal government chose instead to establish a National Board of Youth Service to extend and improve the occupational training activities of the NYA and CCC. The nationally managed institutions took on educational activities allied to occupational training and the result was federally funded secondary schools that competed with the traditional, state and locally funded and oriented, public high schools. According to the EPC, the consequence of this movement was a rigid, class-oriented educational system that quickly abandoned its pursuit of innovation in favor of tight control from the top. Here the EPC gave in fully to its long-held fears that the federal government was going to take over American education and that the consequences of that takeover would be disastrous (Educational Policies Commission, 1944, pp. 2–10).

The rest of *Education for All American Youth* was devoted to outlining what the EPC considered to be an innovative, responsive, state and locally oriented alternative to the federal system of secondary schools. In order to illustrate the need for flexibility, and thereby reinforce the critique of a static, federally managed system, the EPC gave its recommended educational projection in two versions, one for a rural setting and the other for an urban setting. Farmville and American City were the names of the two versions of reform, and they differed subtly in their curricula. What united the rural and urban versions of reform was their agreement on the objectives of the new secondary education. These objectives were enunciated in the report as occupational preparation, citizenship education, the pursuit of happiness, intellectual curiosity and rational thinking, and a democratic ethic (Educational Policies Commission, 1944, p. 21). All but the fourth objective ignored traditional academic content and even that one emphasized intellectual processes over the academic subjects themselves.

In separate chapters devoted to Farmville and American City, the EPC indicated how these objectives were to be reached. A key addition to the high school in both cases was that guidance was to assume a prominent role. Individual interests and differences among students were to be recognized and to be capitalized on as the curriculum was to be customized, through these guidance services, for each student as much as possible. The identification and exploitation of individual differences were the responsibility of the guidance aspect of an invigorated program of psychological services in the new high schools (Educational Policies Commission, 1944, p. 39). In discussing how youth differ, the report listed eight areas: differences occurred in intelligence and aptitude, in occupational interests and outlooks, in the availability of educational facilities, in the types of communities in which they resided, in social and economic status, in parental attitudes and cultural backgrounds, in personal and avocational interests, and in mental health, emotional stability, and physical well-being

(Educational Policies Commission, 1944, pp. 15–16). What the report failed to acknowledge, and what confounded both the idea and the institutionalization of the attempts to acknowledge these differences in a comprehensive high school, was that the relationship among these categories of differences might allow more students to attend the high school at the same time that attendance reified the differences rather than ameliorated them. More specifically, differences in intellectual characteristics might overlap substantially, through heredity, environment, or some combination of the two, with social or class differences, thereby belying the promise of a comprehensive high school as a democratizing agency that enriched the total accomplishments of a society.

While there were differences in the approaches at the two schools in Farmville and American City, they were oriented to the differences in the setting of the institutions rather than to substantively different approaches. Both settings emphasized occupational education as the first priority of the new high school. For example, in Farmville, the tenth graders were enrolled in a course entitled "The World at Work." As part of this experience they would conduct an occupational survey of their community and plan their own futures in accordance with this survey.

The American City discussion was oriented to linking the urban high school with the occupational and vocational needs of the surrounding city. This was to be accomplished through a variety of new commissions in which educators participated with community members in helping the high school become realistically responsive to its community. The institutional goal of all of this activity was a much larger high school with an extremely broadened curriculum that resulted in many more graduates. To accomplish this objective, the report stressed a greatly enriched program of occupational education, including college preparatory education for those suited to this pursuit, along with a comprehensive guidance program to help students choose their desired outcomes. These changes were devoted to meeting specific needs of youth in American City, namely salable skills and allied occupational understandings, health and fitness, democratic citizenship, family comity, intelligent consumerism, scientific understanding, aesthetic appreciation of the arts, wise use of leisure time, respect for others, and rational thought (Educational Policies Commission, 1944, pp. 188–96, 225–26).

Common to both Farmville and American city were the addition of two years to the high school, thus making public education span the thirteenth and fourteenth years or grades. This was to be accomplished through a reorganization of the high school into a seventh, eighth, and ninth grade grouping, a tenth, eleventh, and twelfth grade grouping, and a new institutional grouping made of the two new added years. In a move to increase the reach of the new arrangement at the same time that it was physically detached from the high school, the last two years were to be held in a new community institute, where further occupational study for late adolescents was combined with vocational courses and general educational experiences for adults in the community and the first two years of college education for a few (Educational Policies Commission, 1944, pp. 245–48). In Farmville, where small numbers might preclude a new institutional home for the thirteenth and fourteenth grades, the community institute might have to be attached to the existing high school. Whether the physical setting of the community institute was apart from or adjacent to the high school, the new institution was to be managed by high school administrators, not by higher education officials.

Near the end of the report, the state-level responsibilities for secondary education were addressed through consideration of education in the State of Columbia, in which both Farmville and American city were located. This gave the EPC an opportunity to reiterate several of its long-standing solutions for the public schools. Consolidation of smaller schools, especially in rural districts was one panacea favored by the EPC. Also, the theme of general federal financial aid to the states and local schools, without any control on the part of the national government, was restated. Similarly, the CCC and the NYA again came in for discussion, the theme of which was that, whatever good results these programs had achieved, the time was ripe for their cessation in favor of occupational training in the newly extended and enriched secondary schools proposed in the report. The final statement in the report suspected the authenticity of any national interest in American youth that was not accompanied by a wide-ranging financial commitment to these young people.

Response to the Report

The EPC itself was generally well pleased with the reception of *Education for All American Youth*. By the year after publication, ten regional conferences had been held about the report throughout the country. Twenty thousand copies of the report were distributed in the first year after the report and eighteen thousand copies of an eight-page summary were disseminated (Proceedings, EPC, March 27–29, 1945, pp. 11–13).

The postwar years saw the group move on to new topics, particularly as the Cold War loomed as a significant problem both for America and its schools. Each year, however, the group was given some indication of how successful its volume on the new high school had been. By 1949, "57,512 copies" of *Education for All American Youth* had been distributed, making it the Commission's all-time best seller (Proceedings EPC, October 6–8, 1949, p. 672).

Unaddressed by this façade of satisfaction with the expanded high school, however, were two indications of the difficulties that would plague the EPC reform plans for the American high school. Two attacks on the

EPC proposals for secondary education came respectively from the political Left and from educational traditionalists. The attack from the Left was political, and had been lodged against the EPC since the early 1940s. Though this attack was not devoted to the actual nuts and bolts of the reform, it pointed to a long-standing problem that had plagued, and would continue to frustrate, the National Education Association, parent of the Educational Policies Commission. The issue of concern here was the federal government and its educational activities, specifically the NYA and CCC. The attack from the Right dealt specifically with the educational reform proposals of the EPC. It was launched by a variety of subject matter groups in American education, including those representing English and the sciences. Let us look closely at each of these attacks and see how the EPC responded.

While the political attack from the Left on the EPC predated the publication of *Education for All American Youth* by a few years, *Education for All American Youth* attempted to respond to it directly. Both the attack and the response are worthy of attention here. In December of 1941 and January of 1942, the journal *Frontiers of Democracy*, a leading agency of the social reconstructionist wing of the progressive education reform movement, devoted substantial attention to the EPC's publication *The CCC, The NYA, and the Public Schools* (Bowers, 1969). In two articles in each of the two issues, the authors took issue with the EPC's desire to abolish the CCC and NYA and to replace their activities with an expanded vocational program in state and locally controlled public schools. These two proposals, as noted earlier, were important parts of the EPC program in *Education for All American Youth*. The critics' arguments were several but hinged on two main points. First, the NYA and the CCC had responded to very real economic as well as educational crises among youth in the nation and the federal agencies response included both vocational education and real work activities that paid youth for their efforts (*Frontiers of Democracy* December 15, 1941; *Frontiers of Democracy*, January 15, 1942). That is, the NYA and CCC were performing a significant social and economic amelioration as well as offering educational reforms. Switching these activities to public schools was not advisable, given that the public schools, especially the secondary schools, were bereft of meaningful vocational programs and had never considered issues such as pay for needy students for their labors. Further, handing over these programs to state and local schools was likely to result in their demise, since Congress, which had funded the NYA and CCC, was unlikely to transfer that funding to state educational agencies and local school districts that were beyond its administrative and political reach.

There is no evidence in the Proceedings of the EPC that its members read or otherwise paid any attention to these criticisms. The publication of *Education for All American Youth* in 1944 did respond in part to one of the critics' concerns, however. The proposals for substantial vocational programs in both of its models of the new secondary education, in Farmville and in American City, meant that the EPC was at least aware of what had been accomplished educationally in the two federal agencies. There was no discussion by the EPC of economic issues such as poverty or of paying students in vocational programs, however. Just as importantly, there was little in the way of a concrete proposal to get the federal government to increase the funding for secondary schools to the point that they would be able to launch vocational programs, educational and occupational, that might meet the void created by the demise of the NYA and the CCC. Rather, the EPC continued its strident rhetorical criticism of the two federal educational agencies, its equally strident but substantively empty campaign for federal aid to state and local schools, without any strings, and its reiteration of its belief that these two were enough to win a substantive change in existing federal policy. Of course, another important thing ignored by the EPC was that, given the increased demands on the federal budget to meet the needs of waging a massive war in two distant theaters, it was highly unlikely that any new expenditures in nonwar-related areas would be undertaken. In short, the EPC seemed content to see the CCC and the NYA die, along with the social and economic welfare concerns addressed by policies such as paying students, in favor of replacement by school-based vocational programs that would come in a future time when the professional educators' campaign for federal aid might be successful.

The attack on the EPC from the educational Right was couched directly in response to *Education for All American Youth*. It was voiced in many of the subject matter areas included in the traditional high school curriculum, which came in for heavy revision by the EPC. The sciences were one of the areas that objected and the nature of the objections was spelled out in an article by Oscar Riddle, a biologist doing genetic research at the Carnegie Institution. Riddle originally published his commentary in the journal *Science*, but it was quickly reprinted in an educational outlet (Riddle, February 24, 1945).

Riddle began with praise for certain aspects of *Education for All American Youth* such as its support for federal aid to the schools and its educational plans for young people in the first two years after high school. When it came to curricular issues, however, he was not inclined to praise. His criticism was that in both Farmville and American City, the one, and only, scientific experience in the curriculum from the tenth grade through the two post-high school years was one course in the tenth grade titled "The Scientific View of the World and of Man." Riddle objected for several reasons. One course was surely insufficient to achieve any reasonable understanding of science by high school youth. This insufficiency was highlighted by comparing the one course, unfavorably, to courses in health and physical education that occurred in each year of the high school. The one course in science was not necessarily to be taught by a scientist and it provided no laboratory experience for students, an essential need in a modernizing society that was encountering substantial scientific and technological changes. While agreeing with the EPC that health education was important and should be featured in the high school, Riddle argued that health education needed to be undergirded by solid scientific study and was not a replacement for it. Riddle combined his criticism of the EPC curriculum with a specific reference to wartime conditions near the end of his article.

> If the combined experience of the EPC suggests that the minds of American youth *cannot* be kept or made alert to the facts, principles, and vistas of science, that is highly important; and they should say so. For, to them and to all, it must be clear and certain that the youth cannot escape repeated and important decisions on complex social adjustments imposed by scientific discovery. Following and during a war experience in which our personal safety and our national success alike have been found to depend so greatly upon widespread instruction in the several sciences, it is a paradox that we meet—head on, and even before that emergency passes—a full-scale plan that subordinates the teaching of the sciences in our high schools (Riddle, February 24, 1945, pp. 113–16).

Riddle's academic and disciplinary objections repeated the earlier mentioned concern voiced within the EPC itself about the lack of formal English or other subject matter instruction in the curriculum. His views were also echoed by high school English teachers and biology teachers, and even by industrial arts teachers (Proceedings, EPC, March 27–29, 1945). This last group, which might have likely supported the new vocational thrust, was struggling to maintain the integrity of its own field of study as a legitimate instructional option within the high school curriculum, rather than prostitute it to a practical, vocational orientation.

The academic reaction to the proposals of the EPC received a distinct boost with the publication of the Harvard University faculty report, the famous Harvard Redbook, titled *General Education in a Free Society*, in 1945. Written by a committee of the Harvard College faculty, this report attempted to rethink the academic disciplines, especially in the high school but also in college, to make them speak to the changed conditions of the twentieth century. These changes included two world wars and the consequences of the development of science and technology (Harvard University Committee on the Objectives of General Education in a Free Society, 1945). While there is neither time nor space here to discuss the specifics of the Harvard Redbook in any detail, it is appropriate to note that the Harvard Committee grappled far more seriously with the issues of educational constancy and change than the EPC, which simply took its own priorities for granted and ignored the alternative of academic renovation of the high school in favor of its own program of curricular expansion. The EPC simply asserted its commitment to high school expansion through vocationalism and other curricular additions, without really arguing for it in a serious educational way. The Harvard group, on the other hand, acknowledged that changed conditions demanded changes in curriculum, but specified what needed change in the curriculum and argued for how that change might occur. This was a marked contrast to the EPC assertion, and reassertion, of its predilections and preferences.

At its meeting in December 1945, the EPC took note of the Harvard report, the attention it was getting, and the potential negative reaction it could provoke to *Education for All American Youth*. Rather than undertake a full encounter with the points raised in the Harvard Redbook, the EPC decided not to criticize the Harvard document, but to point out the areas of agreement in the two reports. This was faithfully done in an article on the two reports printed in *The School Review* in February 1946. While the authorship of this article is not specified, it did precisely what the EPC wanted done: it praised the Harvard Report and argued that it and the EPC report were "complementary" rather than opposed to each other. This allowed readers to ignore the serious curricular issues raised by the Harvard faculty and also to ignore the very real clash of ideas in the two reports (Proceedings, EPC December 14–16, 1945; General Education, The Harvard Report and the Educational Policies Commission, February 1946).

As stated earlier, the advocacy of vocationalism and curricular expansionism represented in the EPC proposal was much less problematic than the EPC's reluctance, when engaged by critics, either internally or externally, to respond to the criticism with a principled defense of its proposals and an explication of the rationale behind them. If one cares to call the EPC report a version of the educational progressivism that was dominant within the educational profession, one must also note that this version of educational progressivism was simply stated and restated, rather than substantively argued for, in the report. Similarly the tendency of the EPC in dealing with opposition to its position was to mollify or to relativize criticism by deflecting it, to act as if serious issues did not lie behind the criticism, and to attempt thereby to minimize the very real educational issues that were at stake.

Conclusion

Significantly, *Education for All American Youth* by concentrating on proposals for educational practice and deemphasizing educational principles, represented a different approach for the EPC than the one that engaged the group at its founding. Then, the public schools, beset by ideological critics, by economic calamity, and by elite groups interested in turning both of these to their direct economic and political advantage, received a significant boost from EPC reports that ringingly endorsed the institution of public education and unmasked much of the hidden agenda of the critics. Moving into the practical represented a change in emphasis for the EPC, one dictated, in part at least, by its links to the teachers and administrators represented by the NEA and the AASA (Proceedings, EPC, September 27–29, 1944, pp. 13–44).[3] Yet the very makeup and format of the EPC, a blue-ribbon group that met a few times a year and that considered the largest of issues of relevance to the American educational enterprise, meant that it was not a group likely to be able to develop, sustain, and defend intelligently a curricular program for the American public schools. This, however, lends more strength to the criticism that the EPC might have responded more fully, carefully, and substantively to the ideas raised by its critics, including the Harvard faculty.

Instead, the EPC chose the route of mollification, of obfuscation of real differences, and of procession blithely along its chosen road of becoming a leader for the educational profession and a voice for its practitioners without addressing the issues as seriously as it might have. While there is much food for thought in *Education for All American Youth*, there is also much less there than meets the eye. The opportunistic use of war to conduct a pedagogical agenda that long preceded the war, along with the reluctance to engage critics in a serious discussion of the issues, were both unfortunate and in the long term helped set the stage for the demise of the EPC some years later.

Further, the failure of the EPC to answer the subject-centered critics of *Education for All American Youth* meant that these critics would continue to snipe at the high school, as proffered by professional educators. While the student-centered vision of *Education for American Youth* never was realized in the American high school, the continued affinity for that vision on the part of professional educators caused both them, and their critics, to downplay what actually was going on in the institution in the late twentieth century. This result surely has had something to do with the crisis in which the high school finds itself in the early twenty-first century.

Notes

1. My research on the EPC has been supported by the Spencer Foundation. I gratefully acknowledged that support. This essay was first presented, in different form, at a session at the International Congress of Historical Sciences in Sydney Australia in July of 2005.
2. No publication of the EPC gave credit to an author by name. The Acknowledgment page of this report, however (Educational Policies Commission, 1937), named Charles A. Beard as the primary author.

3. The history of the EPC was punctuated with occasional outbursts of concern for its future on the part of its members, usually sparked by a consideration of whether or not to continue funding on the part of the NEA.

References

Bowers, C.A. (1969). *The progressive educator and the depression: The radical years.* New York: Random House.

Conant, J.B. (1959). *The American high school today: A first report to interested citizens.* New York: McGraw-Hill.

Educational Policies Commission of the Department of the National Education Association and the Department of Superintendence [hereafter Educational Policies Commission] (1937). *The unique function of education in American democracy.* Washington, DC: The Commission.

——— (1940). *Learning the ways of democracy.* Washington, DC: The Commission, 1940.

——— (1941). *The education of free men in American Democracy.* Washington, DC: National Education Association.

——— (1944). *Education for all American youth.* Washington, DC: National Education Association.

Frontiers of Democracy (December 15, 1941), Vol. 8.

——— (January 15, 1942). Vol. 8.

General Education, the Harvard Report and the Educational Policies Commission (1946). *School Review* 54 (February).

Hampel, R. (1986). *The last little citadel: The American high school since 1940.* Boston: Houghton Mifflin.

Harvard University Committee on the Objectives of a General Education in a Free Society (1945). *General education in a free society.* Cambridge, MA: The University.

Hutchins, R.M. (1936). *The higher learning in America.* New Haven: Yale University Press.

Proceedings, Educational Policies Commission (January 10, 1942). National Education Association Archives, Box 935, Washington, DC: National Education Association.

——— (February 19–20, 1942). National Education Association Archives, Box 935. Washington, DC: National Education Association.

——— (June 5, 1942). National Education Association Archives, Box 935. Washington, DC: National Education Association.

——— (March 11, 1944). National Education Association Archives, Box 937. Washington, DC: National Education Association.

——— (March 27–29, 1945). National Education Association Archives, Box 939. Washington, DC: National Education Association.

——— (December 14–16, 1945). National Education Association Archives, Box 940. Washington, DC: National Education Association.

Proceedings, Educational Policies Commission (October 6–8, 1949). National Education Association Archives, Box 945. Washington, DC: National Education Association.

Riddle, O. (1945). "Education for all American youth" from the point of view of a biologist. *School and Society* (February 24).

Urban, W.J. (2000). *Gender, race, and the national education association: Professionalism and Its limitations.* New York: Routledge/Falmer.

Urban, W.J. and Wagoner, Jr., J.L. (2004). *American education: A history.* New York: McGraw-Hill, 3rd ed.

Chapter 4

The Comprehensive High School, Enrollment Expansion, and Inequality: The United States in the Postwar Era

John L. Rury

Among the defining features of the comprehensive high school has been its universal purview. Unlike the narrowly academic or college prep secondary school, it was an institution intended for everyone, as it prepared youth both for vocational careers and their civic roles in a democratic society. The spread of the comprehensive model in American secondary education, in that case, was accompanied by an abiding expectation that all of the nation's teenagers would go to high school, and eventually graduate. This was a widely discussed goal of educational leaders by the middle of the twentieth century (Wraga, 1993; Cremin, 1961). In the decades following World War II, this vision was largely realized in practical terms. This was perhaps the greatest accomplishment of the comprehensive school, even if it was not enough to save it from extensive criticism (Angus and Mirel, 1999).

As a number of studies have noted, the postwar period marked a time of rapid expansion in school participation around the world (Meyer et al., 1977). It was also an era of growth for secondary enrollments in particular. In the United States this was a period when secondary enrollments grew substantially, as American high schools continued to be global leaders in the numbers of students they served (Goldin, 1998; Folger and Nam, 1976). New institutions were established, especially for the nation's black population, and in the rapidly expanding suburban settlements at the edges of major metropolises. Older high schools expanded too, as school systems across the country sought universal secondary enrollment, following the logic of the comprehensive model. This chapter identifies factors associated with changing enrollment patterns in this critical time of educational expansion, when the comprehensive high school arguably achieved its apogee (Rury, 2002).

While high school enrollments were only moderately high in 1950, approaching 60 percent of the relevant age cohort, a solid majority of those enrolled appear to have graduated (Goldin, 1998). Those who did not finish became widely known as the "dropout problem" in the 1950s and 1960s (Brownell, 1954; Dorn, 1996). Again, this term expressed the expectation that secondary graduation would become a commonplace event in the lives of American youth. This was an integral aspect of the comprehensive high school ideal: an institution that would serve every member of society. In the postwar decades this vision approached fulfillment, as the participation rates of teenagers increased substantially, and the dropout rate declined (Campbell, 1966). This chapter assesses this process by examining enrollment growth in American high schools between 1950 and 1970, utilizing state-level data from the U.S. Census.[1]

In certain respects, the experience of American high school students at this time represented a classic case of educational expansion. Enrollment growth for these youth, among the last cohorts to improve their participation in mass schooling to 90 percent in American history, appears to fit the pattern noted by John Meyer and others in describing the worldwide expansion of enrollments in this era (Meyer, Ramirez, and Soysal, 1992; Craig, 1981). Indeed, it might be said that the United States led this process, at least with respect to secondary schools. Meyer and his associates argued that the final stages of expansion were characterized by a "self-generating" process change, largely unaffected by various social and economic factors that had been associated with enrollment growth in previous eras. Of course, there were a number of particular historical circumstances that influenced secondary enrollment in the United States during this period. Groups that traditionally had been denied access to high schools, particularly African Americans, began to gain entry, and the appearance of new suburban communities created additional opportunities for other groups of youth (Baldassare, 1992; Anderson, 1988; Jackson, 1985; Conant, 1961). Thus, this period can be considered a case of educational diffusion under special circumstances; guided by a comprehensive vision of the social and economic role of secondary education, the American experience seems to fit the "self-generating" model of education expansion, but with some caveats.

Research on Enrollment Growth

Although high school attendance became a widespread norm in the postwar period, in earlier times it varied significantly from one region to the next in the United States. At the start of the twentieth century, teenage enrollment was lower in the industrial Northeast, where students often left school before graduation to take jobs in factories. Secondary education was largely restricted to those youth seeking white-collar employment, or preparing for college. Teenage enrollment was high in the West, where industrial employment was less important, and white-collar jobs may have provided an incentive to remain in school. These strong regional trends in school participation were linked to patterns of local economic development and to a variety of historical and cultural factors (Greene and Jacobs, 1992).

A number of studies have examined the growth of high school enrollments, pointing to a range of factors that have influenced it in the past. Many have focused on the changing occupational structure, while others have argued that changing technical requirements in the economy led to higher enrollments across the twentieth century (Walters and O'Connell, 1988; Walters, 1984; Walters, McCammon, and James, 1990; Rubinson and Ralph, 1984). These studies point to a general correspondence between attendance rates and the dictates of the local labor market (Soltow and Stevens, 1977; Fuller, 1983; Rury, 1984).

While much research on attendance has focused on the job market, another set of variables also has been used to analyze enrollment patterns. Studies pursuing this line of inquiry have examined the impact of ethnicity on school attendance, focusing on differences in the experiences of various immigrant and racial groups in American history. In a somewhat similar vein, additional research has considered the influence of social capital as a distinctive community characteristic related in some respects to ethnicity but also reflecting bonds of social cohesion extending beyond particular groups (Rury, 2004). All of these can be considered ethno-cultural explanations of variation in school participation. Most such studies have focused on the nineteenth- and early twentieth century, periods of high immigration, rather than the postwar era, but there is considerable evidence that these factors continued to be important, especially race (Perlmann, 1988; Ralph and Rubinson, 1980; Olneck and Lazerson, 1974; Galenson, 1995). Indeed, it was during the postwar era that racial differences in school participation finally became a critical national concern (Ravitch, 1983).

Finally, yet another interpretive tradition in the study of enrollment growth has been represented by the work of John Meyer and his associates, who have focused primarily on the international expansion of education in the postwar decades. As noted earlier, Meyer has argued that the primary determinants of growth, once enrollments reached a relatively high threshold, are the appeal and demands of the school system itself, and the size of relevant population cohorts to be educated. In short, expansion is seen as largely independent of such exogenous conditions as the economy, employment trends, political development, or cultural differences (Meyer et al., 1977; Ramirez and Meyer, 1980; Meyer, Ramirez, and Soysal, 1992). It thus follows the classic s-shaped curve of a contagion model, wherein the best predictors of growth are prior enrollment levels, especially as participation levels approach unity, or the total number of eligible youth. This explanation of enrollment growth has been demonstrated largely with international data, utilizing cross-sectional analysis of educational expansion in scores of countries (Craig, 1981). This is an especially appropriate analytical frame for assessing the growth of secondary enrollments in the era of the comprehensive high school, given the institution's universalistic goals for popular participation.

High School Enrollment in the Postwar United States

The decades following World War II marked a time of important changes in American education. Enrollments grew as larger numbers of youth completed high school (Goldin, 1998; Folger and Nam, 1976). Geographic differences remained important, however, despite significant advances in the nation's least educationally developed regions. It is possible to see this by examining data on a number of factors as they existed in all 48 states of the continental United States, and in several key regions, displayed in table 4.1.

In 1950, as reported in the census, almost 6 of 10 American youth aged 14 to 17 were enrolled in secondary school. Variation across the 48 continental states was considerable, however, with the lowest enrollment rates in the South and the highest in the West. From this profile, it appears reasonable to suggest that the goal of near-universal (or comprehensive) participation had been achieved in large areas of the country, particularly in the northeast

Table 4.1   Descriptive Statistics: State Means (Weighted by Population)

	National	Northeast	West	South
Percent HS enrol 1950	57.7	65.4	67.7	42.4
Rate of HS enrol exp,	62.4	42.1	39.3	104.6
Percent HS enrol 1970	90.3	92.9	94.2	86.1
Median school yrs	9.44	9.58	11.46	8.41
Percent pop nonwhite	10.5	5.3	5.2	25
Employed percent manufac	25.5	34.3	20.2	18.7
Percent of pop urban	63.8	79.5	75	46.9
Percent pop black 1970	11.3	8.8	5.6	21.1

Notes: All figures calculated from U.S. Census data, 1950 and 1970. Enrollment figures are for youth aged 14–17. Median school years figures are for the adult population (over age 24). 1950 nonwhite population includes African American and Asian population groups. Manufacturing employment and urban population follow census definitions.

(including the Great Lakes region) and on the Pacific coast. As indicated in table 4.1, however, the South exhibited the least advanced profile of educational development in 1950, and it continued to lag in 1970, despite considerable enrollment growth. The South also exhibited the lowest degree of suburban development, and held the vast preponderance of the nation's rural black population. Even with rapid addition of white-collar employment during the postwar era, it was a region that continued to trail the rest of the country on most indices of social and economic development.

In many respects the South posed a challenge to the very idea of the comprehensive high school, one that eventually was overcome. Enrollment levels were one dimension of this, but hardly the most important. The longstanding practice of racial segregation was a far more critical obstacle to the vision of youth from all social strata being educated together in a single institution (Margo, 1990). Following World War II, much of the region's white population remained staunchly resistant to the democratic impulse that lay at the heart of the comprehensive secondary model (Patterson, 2001). Change was difficult, but the numbers indicate that it did occur. The expansion of secondary education across the region demonstrated a gradual acceptance of the notion that high schools should be made available to all members of local communities (Margo and Finegan, 1993; Anderson, 1955; Daniel, 1947). And as Gary Orfield has noted, by the end of the 1960s, Southern states had achieved the nation's highest levels of racial integration in public education. While many white students enrolled in segregated private schools, particularly at the secondary level, public institutions grew both in enrollments and the extent to which they reached students previously excluded (Orfield et al., 1996). In many respects, it was a time of momentous change.

By 1970 the national enrollment profile for secondary students had changed substantially. The country's high school participation rate reached 90 percent, with variation across regions ranging from 86 to 94 percent. Following the pattern of diffusion evident in other studies, growth was slowest in areas with the highest levels at the start of the period, and most rapid in those with low enrollment. As suggested earlier, the South was a geographic focal point of enrollment expansion. Indeed, enrollment rates more than doubled there in just a span of 20 years. As the region with the historically lowest levels of educational development, it clearly had the greatest room for improvement (Cahill and Pieper, 1974). As suggested earlier, access to schools was a critical issue there throughout the postwar period. Enrollment grew more slowly, on the other hand, in the Pacific states and the industrial Northeast, where participation had been relatively high in 1950.[2]

Altogether, in that case, the U.S. experience with high school students in the postwar era appears to have met the conditions associated with the general diffusion model described by Meyer and his associates, with slight exception. As groups of students previously excluded from secondary education enrolled in ever larger numbers, the comprehensive ideal came closer to realization on a national scale. With respect to geographic differences, enrollment expansion appears to have been focused in the South, historically the nation's least educationally advanced region. Association of enrollment with expanding white-collar employment is not generally consistent with the diffusion model, but as indicated in table 4.1, employment growth in this sector was also especially rapid in the South. It is possible, then, that the story of enrollment growth for voluntary students during this time was linked, at least in part, to a peculiar American tale of regional differences in education and economic development. A somewhat more sophisticated test of this thesis is required, however, before it can be considered further.

Postwar Enrollment Growth: A Regression Analysis

To further gauge sources of expansion of secondary enrollment between 1950 and 1970, independent effects of the variables described earlier can be assessed with multiple regression, a statistical technique that permits comparison of various factors in terms of their association with enrollment patterns. This sort of analysis is undertaken later, utilizing state-level data, and proceeds in two steps. The first examines variables related to secondary enrollment levels in 1950, and the second considers how the same factors were related to expansion in high school enrollments, for all 48 states in the continental United States. Since this is not a sample of states, it is not appropriate to refer to statistical significance in connection with this analysis; significance values are identified in tables 4.2 and 4.3 for illustrative purposes only. Several of the variables considered in table 4.1 are included in these analyses, along with some additional factors. By and large, the results that are reported in tables 4.2 and 4.3 appear to support the general diffusion model, which can be linked at least conceptually to the idea of the comprehensive high school.

Each regression is presented in three stages, representing separate but related models comparing factors associated with enrollments. This permits comparison of patterns of association as new factors are added to successive models. The first analysis, in table 4.2, examines high school enrollment levels in 1950. In model one of this analysis, geographic differences in enrollment rates are compared, using dummy variables to represent the regions discussed earlier. As expected, the sign on the South variable is negative and the coefficient is high, as the South lagged considerably behind other regions in secondary enrollment levels at the time. The coefficients for

Table 4.2 Regression, High School Enrollment in 1950, State-Level Data

Model 1		Model 2		Model 3		
Variable	Beta	Variable	Beta	Variable	Beta	Tolerance
Northeast	.104	Northeast	.055	Northeast	.100	.490
West	.180	West	.002	West	.010	.729
South	−.692**	South	−.337**	South	−.317*	.427
		Adult schooling	.357**	Adult schooling	.310*	.358
		Rural blacks	−.284*	Rural blacks	−.300*	.403
				Manufac. employment	−.067	.416
				White collar	.049	.316
Adjusted R/2	.563		.679		.666	

Note: Dependent variable: Percentage of 14–17-year-olds enrolled in high school. State-Level Data, N = 48; * Significant at .05 level; ** Significant at .01 level.

Table 4.3 Regression, Expansion of High School Enrollment, 1950–1970

Model 1		Model 2		Model 3		
Variable	Beta	Variable	Beta	Variable	Beta	Tolerance
Northeast	−.112	Northeast	−.095	Northeast	−.108	.622
West	−.140	West	−.009	West	−.006	.736
South	.684**	South	.426**	South	.393*	.435
		Adult schooling	−.297*	Adult schooling	−.259*	.375
		Rural blacks	.165	Rural blacks	.108	.441
		Suburbs 1970	.033	Suburbs 1970	.066	.675
				Manufac. employment 1970	−.014	.594
				White-collar exp	.186	.436
Adjusted R/2	.533		.571		.569	

Note: Dependent variable: Growth rate of 14–17-year-olds enrolled in high school, 1950–1970. State-Level Data, N = 48; * Significant at .05 level; ** Significant at .01 level.

other regions, the Northeast and Pacific States, are positive, but much weaker. As additional factors were included in models 2 and 3, these basic patterns of association did not change. Other factors associated with high school enrollments in 1950 were general education levels (positive and robust) and the proportion of each state's black population living outside of cities (negative and robust). In general, these patterns of association reflect strong regional differences in enrollment, as high levels of adult education were characteristic of Northern and Western states, and the rural black population was concentrated in the states of the Lower South. The latter factor, of course, was also related to the historic legacy of racial oppression and discrimination in the Deep South, a region that provided relatively few opportunities for African American students to attend secondary schools through almost the entire first half of the twentieth century.

All told, employment patterns appear to have been unrelated to secondary enrollment levels in 1950. The coefficients on both the manufacturing and white-collar employment variables were quite weak and statistically insignificant. This suggests that high levels of manufacturing employment did not represent as important a deterrent to high school enrollment as they had been in the past. Similarly, it indicates that states with relatively large numbers of white-collar workers did not have especially high secondary enrollments, once regional effects and adult education levels were controlled. All told, these patterns of association suggest that secondary enrollment at the start of the postwar period was largely independent of developments in the national economy, and that youth who historically had chosen to work instead of going to school were now enrolled in greater numbers. This was prima fascia evidence of the success of the comprehensive ideal: high schools were now serving a considerably broader segment of society.

If this was the situation in 1950, how did things change in the decades that followed? Table 4.3 presents the results of a similar analysis, with growth of high school enrollments between 1950 and 1970 as the dependent variable, across the 48 continental states. In certain telling respects, the results are nearly the opposite of those reported earlier, and in other respects they are quite similar. The sign on the South regional variable, for instance, is positive and the coefficient robust, indicating that enrollments expanded most rapidly in those states between 1950 and 1970. Similarly, the size of each state's rural black population has a positive sign, but its coefficient is modest in each of the models in table 4.3. The adult education level variable, on the other hand, has a negative sign and was relatively strong, suggesting that enrollments grew the least in states where general education attainment were historically high.[3] In short, expansion of high school enrollments appears to have occurred largely in those parts of the country where they had been lowest in 1950. Generally speaking, the results reported in table 4.3 are consistent with the general diffusion model of enrollment expansion, even if expressed largely in geographic terms.

These models represent rather basic representations of factors associated with high school enrollment levels in this period. Altogether, these variables accounted for less than 60 percent of the state-level variance in American secondary enrollment growth. This means, of course, that a number of additional factors probably also affected secondary enrollment patterns, variables not available in data provided by the decennial census. These regression results, consequently, can be interpreted as providing a partial or incomplete picture of how certain social and economic conditions affected American high schools during the postwar period.

Geography alone, for example, can hardly be expected to account for most of the change in enrollment during these years. Model 3 adds factors related to employment patterns across the period in question: growth in white-collar employment during these decades and the level of manufacturing employment in 1970. As in the analysis of data from 1950, these employment factors did not prove to be as important as prior studies have suggested. Despite the moderate association of white-collar employment with enrollment growth in table 4.3, both employment variables in model 3 were rather modest in their effect, a pattern seen in other cross-sectional studies of educational diffusion (Meyer, Ramirez, and Soysal, 1992). The use of different definitions of these variables did not change this result, suggesting that employment patterns were only weakly related to enrollment growth for these students. Indeed, the r square (or proportion of variance explained) in model 3 was slightly lower than the level reported for model 2, indicating that the inclusion of employment variables failed to provide additional explanatory power to the analysis.

As suggested earlier, these findings suggest a somewhat familiar case of educational diffusion, but under special circumstances. High school enrollment increased by nearly two-thirds in this period across the country, but there was considerable geographic diversity in this process. As the spread of high schools created new norms of educational participation, these students became a part of the secondary population, and this clearly occurred in areas of the country where high school enrollments were relatively low in 1950. Enrollment growth appears to have occurred at the margins of the American social structure: in the South and perhaps among rural blacks. This is a slightly different pattern of educational diffusion than typically discussed in the school expansion literature. The South was a special case in the development of American education, in large part because of its legacy of racist exclusion and discrimination. The postwar period witnessed a direct challenge to that tradition, which may have accounted for some of the rather extraordinary growth in secondary enrollments at the time. It thus appears to have been a distinctive case of diffusion, a process of enrollment growth that partly reflected the vision of the comprehensive high school, an institution that included all the various social groups that comprise the social order and promised to prepare them for citizenship regardless of the vocational paths they may have chosen (Wraga, 1993).

Increasingly, work was something undertaken after education was complete, and fewer youths appear to have left the school to take up employment.

Educational Inequality: Metropolitan Enrollment in 1970

Enrollment rates approached 90 percent for secondary students in the postwar period, but the extent of school participation varied a great deal from one setting to another. As suggested earlier, expansion occurred in areas with low enrollments to start, while areas with high enrollments in 1950 did not experience as much growth. But this does not mean that enrollments were altogether uniform in 1970. As seen in table 4.1, secondary enrollments in the South continued to lag behind the levels observed elsewhere, even if regional differences had narrowed substantially. And there were additional sources of variation in enrollments not evident in the foregoing analysis.

The United States became an increasingly metropolitan nation in the postwar period. By 1970 more than three-quarters of the country's youth lived in cities and suburbs. It was across these settings that the greatest degrees of inequality in educational attainment could be observed. As James Conant famously observed, the form and function of high schools differed a great deal from inner city "slums" to affluent suburban communities (Conant, 1961).

To assess variation in metropolitan high school enrollment, it is necessary to perform a somewhat different analysis. Table 4.4 presents the results of a regression analysis performed with census data from some 122 U.S. metropolitan areas in 1970. Because this is a sample of urban areas, albeit a large and inclusive one, it is also appropriate to discuss the statistical significance (or likelihood of representing true relationships) of various factors in the analysis. The dependent variable is enrollment rates for youth aged 16 and 17, those eligible to leave school voluntarily. Most of the students who reported being in school were enrolled at the secondary level, largely in comprehensive high schools. Because these data were drawn entirely from the 1970 census, some additional independent variables are available for consideration. In particular, it is possible to examine the effect of poverty on enrollment patterns, a question that became especially important among American policy makers in the postwar period, as educators strived to enroll ever larger numbers of students in school.

As in the discussion earlier, the regression analysis in table 4.4 is presented in three stages or models. The first is concerned with regional differences in enrollment across the metropolitan areas being considered, with

Table 4.4 Regression, Metropolitan Teenage Enrollment, 1970

Model 1		Model 2		Model 3		
Variable	Beta	Variable	Beta	Variable	Beta	Tolerance
Northeast	.117	Northeast	.051	Northeast	−.004	.583
West	.102	West	−.017	West	−.056	.685
Southeast	−.396**	Southeast	−.217**	Southeast	−.185*	.727
		Adult schooling	.216**	Adult schooling	.258*	.442
		Black population	−.078	Black population	−.165	.492
		Poverty population	−.403**	Poverty population	−.542**	.416
				Part-time employment	−.231*	.426
				Clerical employment	.044	.605
				Profess employment	−.041	.512
Adjusted R/2	.194		.405		.414	

Note: Dependent variable: Proportion of 16- and 17-year-olds enrolled in school, 1970. Metropolitan Area (SMSA) Data, N = 122; * Significant at .05 level; ** Significant at .01 level.

regions defined in slightly different but broadly parallel terms as earlier. Given the convergence of enrollment patterns across the postwar period, it is telling to find that the Southeast variable in model 1 is strong, significant, and has a negative sign. Indeed, it remains negative and significant in all three models, indicating that the South continued to trail the rest of the country in secondary schooling even in 1970. The educational heritage of the region apparently persisted even in its larger metropolitan areas, and was independent of a variety of other factors included in table 4.4.

Model 2 introduces several factors that might be described as cultural and economic in nature. The first is adult levels of education, the same variable utilized earlier but this time calculated with 1970 data. The second is the portion of each metropolitan area's population that is African American, and the third is the proportion of residents with incomes under the federal poverty level. The first and third of these variables are important and significant, but their signs are different. Not surprisingly, the sign on the adult education variable is positive, and the poverty variable is negative. The black population variable is negative, suggesting that African American youth had lower enrollment rates at these ages, but it failed to achieve significance. Tellingly, poverty is clearly the strongest factor in both models 2 and 3. Even with the rapid expansion of high school enrollment in the years leading up to 1970, the poverty levels across metropolitan areas were a critical source of variation in school participation for 16- and 17-year-old youth. In urban areas with relatively high rates of poverty, significantly higher numbers of these youth had dropped out of school. It is little wonder that social status became such an important focal point of policy initiatives during this period (Baron, 1971).

Finally, model 3 introduces a number of employment variables, only one of which appears to have been important. Clerical and professional employment levels were generally unrelated to high school enrollment for these youth, although the number working while in school was significant and negatively associated with school participation. This suggests that areas with abundant part-time work opportunities provided a context that militated against enrollment, even controlling for poverty, region, and other factors. With the expansion of secondary enrollment during this era, part-time employment became an increasingly important factor in the lives of American youth, one that apparently did not help them to remain in school in many instances. This is a finding corroborated in other studies (Buchman, 1989; Pallas, 1993; Oettinger, 1999).

The analysis presented in table 4.4, in that case, has pointed to persistent patterns of inequality in secondary enrollment levels, even after the spread of the comprehensive ideal had made high school a nearly ubiquitous experience for American youth. The most striking factor in this analysis, of course, was poverty, which exerted a strongly negative pull on enrollments. This finding is hardly a surprise today, following decades of research that has demonstrated the magnitude of this relationship (Jencks et al., 1979; Duncan and Brooks-Gunn, 1997). But it probably was not until the era of universal high school enrollment that its weight was fully appreciated. In the past, large numbers of poor youth left school to join the labor force. During the postwar period this does not seem to have been the case, the growing importance of part-time employment notwithstanding. By and large, the high school had ceased to compete with the labor market, as the comprehensive ideal held that all youth needed to complete secondary education before entering the adult world of full-time employment. Even so, the negative association of poverty and teenage enrollment suggests that a significant portion of the population remained alienated from the institutions that were supposed to represent the comprehensive ideal. This posed a major challenge to the high schools, one widely recognized at the time (Hummel and Nagle, 1973).

Regional differences remained important, especially in the South, and some places provided more or less hospitable environments for enrollment depending upon long-standing educational traditions or evolving employment opportunities for youth. Race undoubtedly was a factor as well, despite its lack of significance in this analysis. Universal high school participation, after all, did not mean that all students stayed enrolled or eventually graduated. As secondary attendance expanded to include larger numbers of the nation's youth, it seems that involvement with schools continued to reflect larger patterns of inequity in American life. This demonstrated the limits of the comprehensive high school's success in the postwar period. Even with its phenomenal growth, the American secondary school ultimately failed to reach a critical segment of the nation's youth.

Conclusion

In the two decades following 1950, high school enrollment became a nearly universal form of behavior among American teenagers. As has been noted by a number of commentators, the United States led the world in secondary education, at least in terms of attendance (Goldin and Katz, 2001). As noted earlier, however, the rate of advancing participation was not the same in all parts of the country. As predicted by prior research on educational expansion, growth was greatest in those parts of the country where enrollments were relatively low at that start of the period. In this respect, the postwar American experience appears to conform to the worldwide pattern of diffusion observed by Meyer and his collaborators some thirty years ago (Meyer et al., 1977). By and large, economic and by extension technological factors do not seem to have been related to the state-level differences in enrollment growth for this age cohort, at least insofar as they have been captured in this analysis. This too is consistent with the findings of prior cross-sectional research on the period (Craig, 1981).

Regional differences in education have long been an important theme in the United States. Because the South had been a historically underdeveloped region, in education along with a number of other domains, it is not surprising to find that overall enrollment rates there grew rapidly during these years (Cahill and Pieper, 1974). It appears, however, that at least some of this advance was due to states with large numbers of African Americans living in rural areas, a group that traditionally had been excluded from secondary education. Yet, even when this is controlled for, overall rates of enrollment in expansion in the South appear to have led the rest of the country. In this respect the South appears to confirm the general expectations of the diffusion model, yet one also has to account for the special role of race in the region's history and its experience during the postwar era. The advent of the Civil Rights movement, and the concomitant expansion of black education, helped to make this a somewhat unusual case of educational diffusion in regional terms (Levitan et al., 1975). In the postwar era, the idea of the comprehensive high school appears to have finally made its mark on this critical region.

Despite the rapid expansion of enrollments, inequality continued to characterize American secondary education at the end of this period. Regional differences persisted, even in the nation's major metropolitan areas, and poverty loomed large as a factor shaping the participation of American youth in the high school, even as it became a universal institution. By 1970, as virtually all Americans went to secondary school for at least a portion of their educational experience, a new set of challenges emerged to bedevil educators' hopes for a future when everyone would graduate from high school. The great promise of the comprehensive high school, after all, was to create an institution that could serve the needs of all Americans, and the inequities that appeared so starkly in 1970 were a grim reminder that yet more work would be necessary before such a pledge could be fulfilled. It was the task of addressing these concerns that would preoccupy policy makers in the years to come, and in many respects theirs is a challenge that has yet to be resolved.

Of course, this essay has primarily been concerned with access to the high school during the postwar era, as reflected in rising enrollment rates. It has had nothing to say about the academic achievement of the students who attended secondary school at the time. Following the attainment of near-universal enrollment levels, public attention turned to the task of achieving universal graduation, a goal that proved far more elusive than merely getting teenagers into school (Angus and Mirel, 1999). The great promise of the comprehensive high school, after all, was one of bringing students from all the various segments of society together. Relatively little was said about what they were supposed to do there, or the circumstances under which they were supposed to finish. As I have pointed out elsewhere (Rury, 2002), the comprehensive model of American secondary education was increasingly called into question when secondary educators turned to these issues in the years after 1970.

By and large, studies have indicated that achievement patterns in secondary education have largely mirrored the many factors discussed earlier with regard to variation in enrollment rates: poverty, race, and broad trends in social inequality. Students from working-class families, racial and ethnic minorities, those living in underdeveloped or economically declining areas have historically exhibited lower levels of achievement than others (Duncan and Brooks Gunn, 1997). In the comprehensive high school it was expected that some students would exhibit higher levels of achievement than others. Following the civil rights struggles over educational equity, however, variation in performance along class and racial lines became a point of controversy. This contributed to a growing tide of calls to alter the comprehensive model of secondary education (Eurich, 1970; NCRSE, 1973). The publication of *A Nation at Risk* in 1983 signaled an abiding national concern for universally high levels of academic achievement, marking a new stage in the history of American school reform (Angus and Mirel, 1999; NCEE, 1983). With this, the accomplishments of the comprehensive high school, particularly the attainment of near-universal enrollment, faded in significance. And it was these more recent historical circumstances that have set into motion the current dilemmas facing the American secondary education.

Notes

1. All data in this study are drawn from the published reports issued by the United States Census for 1950 and 1970. Data for individual states were taken from the state volumes for each year. See *United States census of population, 1950*. Washington, D.C.: U.S. Govt. Print. Off., 1951; United States. Bureau of the Census. *1970 census of population*. Washington, D.C.: U.S. Bureau of the Census, 1971. I am indebted to Sylvia and Nick Martinez, who performed the bulk of data collection for this analysis at the University of Kansas; city-level data were collected at DePaul University by Aaron and Derek Rury and Neil Sullivan.
2. Several aspects of table 4.1 bear explanation. All figures are means of state-level data weighted by population size. Rate of high school expansion reflects not only increased enrollment, but population growth as well, which was considerable due to the "baby boom" of the 1950s and early 1960s. The rural black population figure is the percent of the state's black population living outside of urban areas. The suburban population was defined as "urban fringe population" in the 1970 census. White-collar employment is a combination of professional, managerial, and clerical employment categories in the census, for both men and women, expressed as a proportion of the total labor force. Operative employment is the number of "factory operatives" expressed as a share of the total labor force, for men and women.
3. Typically, enrollment growth is modeled by using participation levels at the start of the period in question as a baseline. Using 1950 high school enrollments as an independent variable in model 3 did not change the other factors in the equation; it just produced a slightly higher tolerance coefficient and VIF values. As indicated in table 4.2, adult education levels were positively associated with teenage enrollment levels in 1950. Both factors are potent indicators of school participation levels at the start of the period, correlated at 0.71 for the 48 contiguous U.S. states. For a widely influential statistical analysis of enrollment growth, see Meyer et al. (1977).

References

Anderson, C.A. (1955). Inequalities in schooling in the South. *The American Journal of Sociology* 60 (6) (May): 547–61.

Anderson, J.D. (1988). *The education of blacks in the South, 1860–1935*. Chapel Hill: University of North Carolina Press.

Angus, D.L and Mirel, J.E. (1999). *The failed promise of the American High school, 1890–1995* New York: Teachers College Press.

Baldassare, M. (1992). Suburban communities. *Annual Review of Sociology* 18 (1992): 475–94.

Baron, H.M. (1971). Race and status in school spending: Chicago, 1961–1966. *The Journal of Human Resources* 6 (1) (Winter): 3–24.

Brownell, S.M. (1954). Unsolved problems in American education. *The School Review* 62 (9) (December): 519–26.
Buchman, M. (1989). *The script of life in modern society: Entry into adulthood in a changing world* (p. 132). Chicago: University of Chicago Press.
Cahill, E.E. and Pieper, H. (1974). Closing the educational gap: The South versus the United States. *Phylon* 35 (1) (1st Qtr.): 45–53.
Campbell, G.V. (1966). A review of the drop out problem. *Peabody Journal of Education* 44 (2) (September): 102–09.
Conant, J.B. (1961). *Slums and suburbs: A Commentary on schools in metropolitan areas.* New York: McGraw-Hill.
Craig, J.E. (1981). The expansion of education. *Review of Research in Education* 9: 151–213.
Cremin, L. (1961). *The transformation of the school: Progressivism in American education, 1870–1957.* New York: Harper & Row.
Daniel, W.G. (1947). Availability of education for Negroes in the secondary school. *The Journal of Negro Education* 16 (3) (Summer 1947): 450–58.
Dorn, S. (1996). *Creating the dropout: An institutional and social history of school failure* Westport, CT: Praeger Publishers.
Duncan, G. and Brooks-Gunn, J.E. (Eds.) (1997). *The social consequences of growing up poor.* New York: Russell Sage Foundation.
Eurich, A.C. (Ed.) (1970). *High School 1980; the shape of the future in American secondary education.* New York, Pitman Publishing.
Folger, J.K. and Nam, C.B. (1976). *Education of the American population.* New York: Arno Press.
Fuller, B. (1983).Youth job structure and school enrollment, 1890–1920. *Sociology of Education* 56 (3) (July): 145–56.
Galenson, D.W. (1995). Educational opportunity on the urban frontier: Nativity, wealth, and school attendance in early Chicago. *Economic Development and Cultural Change* 43 (3) (April): 551–63.
Goldin, C. (1998). America's graduation from high school: The evolution and spread of secondary schooling in the twentieth century. *The Journal of Economic History* 58 (2) (June): 345–74.
Goldin, C. and Katz, L.F. (1999). Human capital and social capital: The rise of secondary schooling in America, 1910–1940. *Journal of Interdisciplinary History* 29 (4) (Spring): 683–723.
——— (2001). The legacy of U.S. educational leadership: Notes on distribution and economic growth in the 20th century. *The American Economic Review* 91 (2) (May 2001): 18–23.
Greene, M.E. and Jacobs, J.A. (1992). Urban enrollments and the growth of schooling: Evidence from the U. S. 1910 census public use sample. *American Journal of Education* 101 (1) (November): 29–59.
Hummel, R.C. and Nagle, J.M. (1973). *Urban education in America; problems and prospects.* New York: Oxford University Press.
Jackson, K.T. (1985). *Crabgrass frontier: The suburbanization of the United States.* New York: Oxford University Press.
Jencks, C.R. and Bartlett, S. (1979). *Who gets ahead?: The determinants of economic success in America.* New York: Basic Books.

Levitan, S., Johnston, W.B., and Taggert, R. (1975). *Still a dream: The changing status of blacks since 1960*. Cambridge, MA: Harvard University Press.

Margo, R.A. (1990). *Race and schooling in the South, 1880–1950: An economic history*. Chicago: University of Chicago Press.

Margo, R.A. and Finegan, T.A. (1993). The decline in black teenage labor-force participation in the South, 1900–1970: The role of schooling. *The American Economic Review* 83 (1) (March): 234–47.

Meyer, J.W., Ramirez, F.O., Rubinson, R., and Boli-Bennett, J. (1977). The world educational revolution, 1950–1970. *Sociology of Education* 50 (4) (October): 242–58.

Meyer, J.W., Ramirez, F.O., and Soysal, Y.N. (1992). World expansion of mass education, 1870–1980. *Sociology of Education* 65 (2) (April): 128–49.

National Commission on Excellence in Education (NCEE) (1983). *A Nation at risk: The imperative for educational reform*. Washington, DC: Government Printing Office.

National Commission on Reform of Secondary Education (NCRSE) (1973). *The reform of secondary education: A report to the public and the profession*. New York: McGraw Hill.

Oettinger, G.S. (1999). Does high school employment affect high school academic performance? *Industrial and Labor Relations Review* 53 (1) (October): 136–51.

Olneck, M.R. and Lazerson, M. (1974). The school achievement of immigrant children: 1900–1930. *History of Education Quarterly* 14 (4) (Winter): 453–82.

Orfield, G., Eaton, S.E., and the Harvard Project on School Desegregation (1996). *Dismantling desegregation: The quiet reversal of Brown V. board of education*. New York: Norton.

Pallas A.M. (1993). Schooling in the course of human lives: The social context of education and the transition to adulthood in industrial society. *Review of Educational Research* 63 (4) (Winter): 409–47 (423).

Patterson, J.T. (2001). *Brown v. board of education: A civil rights milestone and its troubled legacy*. New York: Oxford University Press.

Perlmann, J. (1988). *Ethnic differences: Schooling and social structure among the Irish, Italians, Jews, and Blacks in an American city, 1880–1935*. New York: Cambridge University Press.

Ralph, J.H. and Rubinson, R. (1980). Immigration and the expansion of schooling in the United States, 1890–1970. *American Sociological Review* 45 (6) (December): 943–54.

Ramirez, F.O. and Meyer, J.W. (1980). Comparative education: The social construction of the modern world system. *Annual Review of Sociology* 6 (1980): 369–99.

Ravitch, D. (1983). *The troubled crusade: American education, 1945–1980*. New York: Basic Books.

Rubinson, R. and Ralph, J. (1984). Technical change and the expansion of schooling in the United States, 1890–1970. *Sociology of Education* 57 (3) (July): 134–52.

Rury, J. (1984). Urban structure and school participation: Immigrant Women in 1900. *Social Science History* 8 (3) (Summer): 219–41.

——— (2002). Democracy's high school? Social change and American secondary education in the post-Conant era. *American Educational Research Journal* 39 (2) (Summer): 307–36.

Rury, J. (2004). Social capital and secondary education: Inter-urban differences in teenage enrollment rates in 1950. *American Journal of Education* (August): 293–320.

Soltow, L. and Stevens, E. (1977). Economic aspects of school participation in mid-nineteenth-century United States. *Journal of Interdisciplinary History* 8 (2) (Autumn): 221–43.

Walters, P.B. (1984). Occupational and labor market effects on secondary and post-secondary educational expansion in the United States: 1922 to 1979. *American Sociological Review* 49 (5) (October): 659–71.

Walters, P.B. and O'Connell, P.J. (1988). The family economy, work, and educational participation in the United States, 1890–1940. *The American Journal of Sociology* 93 (5) (March): 1116–152.

Walters, P.B. and James, D.R. (1992). Schooling for some: Child labor and school enrollment of black and white children in the early twentieth-century South. *American Sociological Review* 57 (5) (October): 635–50.

Walters, P.B., McCammon, H.J., and James, D.R. (1990). Schooling or working? Public education, racial politics, and the organization of production in 1910. *Sociology of Education* 63 (1) (January): 1–26.

Wraga, W. (1993). *Democracy's high school*. Washington: University Press of America.

Contemporary Perspectives

Chapter 5

Breathing Life into Small School Reform: Advocating for Critical Care in Small Schools of Color

*Rene Antrop González and
Anthony De Jesús**

Introduction

We unequivocally declare death on the social institution known as the large comprehensive urban high school, because it has miserably failed students of color, particularly Latina/o youth. This horrific circumstance manifests itself in various traditional urban teaching and learning structures that facilitate inevitable failure. These structures are characterized by overcrowded classrooms that do not allow for the important establishment of meaningful, high-quality relationships between students and teachers, watered down curricula that privilege rote memorization, decontextualized skill sets, the overall numbing of the mind, and low academic expectations that reflect the disturbing notion that students of color are inferior and have nothing to offer our nation. As consequences of these institutional barriers, high school non-completion rates are astronomically high, such that 58 percent of Latino youth between the ages of 18 and 24 are pushed out of high school (Rodríguez, Antrop-González, and Reyes, 2006). Thus, it is imperative that educational stakeholders such as students, parents/caregivers, community members, school-based institutional agents, and scholars begin to rethink how we conduct the very important endeavor of teaching and learning in urban schools.

Scholars have explained the various ways in which the majority of urban school curricula are inherently designed to privilege white, middle-class values and ignore or dehumanize the sociopolitical, historical, and linguistic realities of urban youth of color (Nieto, 1998, 2000; Valenzuela, 1999). The process of subtractive schooling is one in which schools "divest these [Latino] youth of important social and cultural resources, leaving them progressively vulnerable to academic failure" (Valenzuela, 1999, p. 3) and is manifested through the absence of culturally relevant pedagogy (Ladson-Billings, 1995) that honors the lives of Latina/o students, their families, and communities. This absence of culturally relevant pedagogy reveals itself in multiple ways, such as through the lack of reading materials in Spanish and/or English that portray Latino life in the United States, through school programs that deny the importance of bilingual/bicultural education, through school staff that hold their Latino students to low academic expectations and are simply not interested in establishing nor maintaining interpersonal relationships with their students. Moreover, it is very common to find a great lack of bilingual/bicultural teachers, staff, and administrators at many traditional, large traditional high schools. Consequently, in response to the failure of comprehensive high schools to offer youth of color excellent educational opportunities, numerous urban centers, such as Chicago, Cleveland, Detroit, Milwaukee, New York, and Portland have looked to small school reform as a plausible answer.

Hence, this chapter presents our respective research as Puerto Rican/DiaspoRican[1] scholars concerning the history, curricula, and student experiences of two community-driven small high schools that have served as alternatives to failing comprehensive high schooling in their respective U.S.-based Puerto Rican/Latino communities—the Dr. Pedro Albizu Campos High School (PACHS) in Chicago, Illinois and El Puente Academy for Peace and Justice (El Puente) in Brooklyn, New York. As small community-based schools, lessons from PACHS and El Puente promise to inform the debate on the nature of small school reform and its relationship with marginalized communities. We strongly feel that the proliferation of these particular types of community-based small schools have the potential to dismantle failing comprehensive high schools as we know them. Nonetheless, we must also be wary of those small schools that only reproduce the oppressive structures of comprehensive high schools and reveal a general disregard for social justice and meeting the needs of urban youth of color (Fine, 2000; Mohr, 2000). Hence, at a time when the small schools movement has taken hold in major cities throughout the United States, we believe there are important lessons to learn from the experiences of these two Latino community schools.

Moreover, we believe these specific types of community-driven small schools will go much farther to yield the overwhelming sense of racial/ethic and linguistic alienation and poor academic achievement that most urban youth of color experience in comprehensive high schools. Our analyses, informed these youth, emphasize that educational projects like PACHS and

El Puente are not created in sociopolitical/historical vacuums but emerge organically from communities of color and their respective struggles for improved educational opportunities as well as political movements for self-determination, community control, and decolonization. In turn, the formal and informal curricula of these schools reflect the cultural values and political realities of the communities that established them and, we argue, provide students with educational and social experiences closely aligned with their community and cultural resources or *funds of knowledge* (Moll et al., 1992) and more fully embody the educational interests of Latina/o communities. To the extent possible, these schools reflect the notion of schools being created *for the community, by the community*. Decades before philanthropy and school districts signed onto the small schools movement, these communities established small schools with few material, economic, and human resources as cultural and community centers to address the fact that their children were being left behind by large, impersonal, and culturally hostile public schools.

Ultimately, through our conversations with youth and school staff, we reveal that the combination of high-quality interpersonal relationships, high academic expectations, and mentorship is crucial for engaging students in the learning process and fomenting academic success. Thus, we interpret this combination of instrumental relationships (Stanton-Salazar, 2001) and high academic expectations (Katz, 1999) through the scholarship on caring in education and its intersections with the social and cultural capital literature and link these notions to rich descriptive data from our experiences at PACHS and El Puente.

Theoretical Perspectives on Caring in Education

A number of theorists (Dance, 2002; Katz, 1999; Noddings, 1984, 1992; Thompson, 1998; Valenzuela, 1999) have argued that experiences of caring within the student/teacher relationship are essential to student engagement and suggest that the educational success of marginalized students in particular often hinges on "being engaged in a caring relationship with an adult at school" (Valenzuela, 1999). Contrary to racially uncritical notions of caring emerging from white feminists, Thompson (2002) advances a strong critique of the colorblind assumption in white feminist notions of caring as an emotion-laden practice characterized by low expectations, which are motivated by taking pity on students' social circumstances (Katz, 1999). Within the Puerto Rican experience we call this the "*Ay Bendito* syndrome" referring to the Spanish-language exclamation of pity. We refer to this form as *soft* caring because it is characterized by a teacher's feeling sorry for a student's circumstances and lowering their academic expectations of them out of pity. While rooted in a legitimate expression of concern for the well-being of another, we consider the actions and analysis emanating from an emotional response as more important than well-intentioned attempts to care for youth of color.

Alternatively, we argue that communities of color understand caring within their sociocultural, gendered, and economic contexts and believe that caring has traditionally existed within differential economic contours for disenfranchised communities—particularly within the experience of black communities, as Thompson (1998) points out:

> Whereas caring in the White tradition is largely voluntary emotional labor performed in an intimate setting or else underpaid work in a pink-collar profession like teaching or nursing, caring in the Black community is as much a public undertaking as it is a private or semi-private concern. It is not surprising, therefore, that caring in the Black community is not understood as compensatory work meant to remedy the shortcomings of justice, as in the "haven in a heartless world" model. (p. 9)

The scholar who has made the greatest contribution to date in exploring the ways in which identity and context shape experiences of caring for Latina/o students is Angela Valenzuela (1999) who, in her book *Subtractive Schooling: US Mexican Students and the Politics of Caring*, describes the ways in which traditional urban comprehensive high schools (like Seguín High School in her study) are organized formally and informally in ways that divest Latina/o students of "important social and cultural resources, leaving them progressively vulnerable to academic failure" (p. 3). She goes on to observe that "rather than building on students' cultural and linguistic knowledge and heritage to create biculturally and bilingually competent youth, schools subtract these identifications from them to their social and academic detriment" (p. 25). Building on Noddings's (1984, 1994) caring framework, Valenzuela (1999) analyzes competing notions of caring (aesthetic vs. authentic) among teachers and students that are rooted in fundamentally different cultural and class-based expectations about the nature of schooling. These expectations inevitably clash, and when they do, fuel conflict and power struggles between teachers and students who see each other as not caring. As Valenzuela (1999) observes:

> The predominately non-Latino teaching staff sees students as not sufficiently *caring about* school, while students see teachers as not sufficiently *caring for* them. Teachers expect students to demonstrate caring about schooling with an abstract, or *aesthetic* commitment to ideas or practices that purportedly lead to achievement. Immigrant and U.S.-born youth, on the other hand, are committed to an *authentic* form of caring that emphasizes relations of reciprocity between teachers and students. (p. 61)

To succeed academically at Seguín and many other public comprehensive high schools in the United States, students must conform to the faculty's value of aesthetic caring, "whose essence lies in an attention to things and ideas" (Valenzuela, 1999, p. 22). While "attention to things and ideas" are an important element of academic learning and development, Latina/o students often resist this notion because they experience the cultural and social distance between them and their teachers as depersonalizing and inauthentic. As a result, Valenzuela (1999) suggests that "(c)onceptualizations of educational 'caring' must more explicitly challenge the notion that assimilation is a neutral process so that cultural and language affirming curricula may be set into motion" (p. 25). Additionally, this analysis suggests that mediating the tensions between aesthetic and authentic caring is related to school structures that emphasize or delimit particular forms of teacher caring.

As scholars of color, we agree with Thompson (2002) that notions of educational caring are not colorblind or powerblind and that communities of color necessarily understand caring within their sociocultural context. This context must be acknowledged in order to forge a new caring framework that privileges the cultural values and political economy of communities of color as a foundation for education. This premise is at the heart of our conceptualization of *critical care*, a term that captures the ways in which communities of color may care about and educate their own and their intentions in doing so. In addition, we also feel that white educators have the potential to become allies of urban youth of color only if they are willing to seriously reflect on the extent to which they are willing to adopt critical care in their work with their students.

In light of critiques of white feminist conceptions of caring, McKamey (2002) suggests that a new theoretical conversation must begin with what she calls a *process theory* of caring that deepens the conversation and "provides insight to the potential complexities and contradictions inherent within caring interactions, interpretations, expressions, and contexts" (p. 39). The notion of *critical care*, as described herein, represents our contribution to deepening this theoretical conversation.

As Latino scholars seeking to account for the inherent intersections between caring, identity, and context, the purposes of this chapter are to forward a theory of critical care based on our research and integrate Thompson's critique of colorblind forms of caring with understandings of social and cultural capital. The interests of communities of color, specifically Latina/o communities, are translated into school cultures and practices aimed at engaging students in learning linked to broader goals of community survival and development. We are particularly interested in the ways in which Latina/o *funds of knowledge* (Moll et al., 1992) may constitute formal and informal curricular and pedagogical practices leading to the transformation of educational outcomes for marginalized students.

The student voices presented in this chapter describe the ways we believe PACHS and El Puente have created culturally additive learning communities underscored by high-quality relationships and high academic expectations that reflect an ethic of *critical care* and illustrate the practice of *hard* caring in which power is shared with students rather than over them (Kreisberg, 1992). This particular form of caring is characterized by high academic expectations. While the small size of El Puente and PACHS provide an important context for authentically caring relationships to occur, our analysis reveals that it is the cultural and interpersonal substance of the formal and informal curricula at the two schools that lead to important outcomes, such as students' sense of belonging and high academic achievement.

Building on the educational caring scholarship (McKamey, 2002; Thompson, 2002; Valenzuela, 1999) we believe that educational projects, like PACHS and El Puente, explicitly acknowledge community and student contexts and seek to affirm the identities, social and cultural resources of Latina/o students and constitute the best possible response to traditional forms of non-caring, subtractive schooling and the systematic failure these produce. The voices of students we interviewed support these theoretical claims. In the following section, we provide brief sociohistorical and political descriptions of the two schools that inform our analyses.

Methods

We employed ethnographic, case study methods in order to make sense of and describe the schooling experiences of students, teachers, and other staff members at each school. These methods consisted of structured individual interviews, focus groups, and participant observations. In addition, we collected school-related historical and curricular documents. These documents included brochures, archived newspaper reports, and a copy of the formal curriculum. These documents enabled us to learn more about each school's history and operations, such as why it was founded, how it was funded and accredited, how the school was operated administratively, and how its curricula were structured. We analyzed these data in order to isolate recurring themes and, our analyses were guided by the following areas of inquiry:

(1) What types of interactions took place between students and teachers?
(2) How did students, teachers, and administrators describe their respective experiences at the high school?
(3) Why did students choose to attend these particular high schools rather than any other of the traditional public high schools?
(4) How similar and/or different were the experiences of the female and male participants?
(5) Were the experiences of the school's Puerto Rican students and staff members and the school's non-Puerto Rican students and staff members different and/or similar (e.g., African Americans, Mexican Americans, those of multiple Latino ethnicity, and/or white)?

El Puente Academy for Peace and Justice (El Puente): History and Context

El Puente Academy for Peace and Justice is a small high school in New York City that emerged from (and is part of) El Puente, a community based organization in Brooklyn, New York that was founded as an after-school and cultural arts center in 1982 by Puerto Rican and Latina/o activists in Williamsburg/Los Sures. A historically poor and working-class Latino/a community, El Puente today is a vibrant institution that incorporates the Academy, three youth development centers (after-school programs), and a number of other community development initiatives. The organization was initially founded in response to a protracted period of youth violence during the late 1970s and early 1980s and the inability of existing social service agencies and schools to address these problems. Eastern District High School, the zone school for Williamsburg, symbolized for El Puente's founders what was wrong with the Board of Education.[2] The conditions Valenzuela (1999) describes as *subtractive schooling* resonate strongly with the educational experiences of Latina/o youth in North Brooklyn and throughout New York City at the time El Puente was founded. In a 1988 *New York Times* article documenting dropout rates in New York City, Luis Garden-Acosta, El Puente's principal founder, summarized the messages he believed Latino youth received as part of their education in New York City public schools. His statement sheds light on the way El Puente Academy's founders defined the educational problems for the Latina/o students they later would seek to address through their own school: "Young people are being given a message: Your culture is not good enough; your language stinks; you have to adjust to our culture, it's an insensitive cultural response by the Board of Education and the educational system in general" (Carmody, 1988).

In response to these conditions, the founders of El Puente sought to create a holistic after-school learning community that affirmed the language, culture, and identities of Latino/a students and linked the individual development of students to a broader vision of community development. In their efforts to develop effective youth development and culturally responsive after-school programming based on principles of peace and justice and human rights, El Puente's founders identified the need to address the schooling of young people in their community. In 1993, El Puente opened as a New York City public high school under the auspices of New Visions for Education, a nonprofit initiative founded "to create a critical mass of small, effective schools that equitably serve the full range of children in New York City" (Rivera and Pedraza, 2000, p. 227). Now in its tenth year, El Puente Academy for Peace and Justice serves 150 students in grades 9 through 12, 87 percent of which are Latina/o and 11 percent African

American. The majority of students are residents of North Brooklyn and come from low-income backgrounds. While now a New York City Public School, the fact that El Puente was founded by Latina/o community activists who explicitly sought to create a school whose purpose is linked to community development (for the community, by the community) creates organizational and instructional conditions that are more reflective of the interests and values of local Latina/o residents than those of professional school district administrators or school planners. In this context, educational caring at El Puente (and PACHS) emerges from more profound origins and takes on additional meaning.

The Dr. Pedro Albizu Campos High School (PACHS): History and Context

The Dr. Pedro Albizu Campos Alternative High School (PACHS) was founded in 1972 as a response to the Eurocentric-based curricula and high dropout rates that Puerto Rican students had been experiencing in Chicago's public high schools. Historically, the dropout rates among Puerto Rican urban high school students in the United States have ranged anywhere between 45 and 65 percent (Flores-González, 2002). An April 8, 1973 article in the *Chicago Tribune* titled, "Puerto Ricans Here Set Up Free School to Aid Dropouts," described the depressing social and pedagogical conditions that led to the founding of this high school:

> The school, which opened in February [1972], is geared to aid Puerto Ricans who have dropped out of Tuley, Wells, and Lake View High Schools. It also serves as an alternative for Puerto Rican students who are considering leaving school because of academic or personal problems. . . . Puerto Rican students, parents, and community leaders have long complained that the Chicago public school system is counterproductive and generally apathetic to the real needs of Puerto Rican students.

Originally named "La Escuela Puertorriqueña (the Puerto Rican School)" the high school was established to address the educational needs of its mostly Puerto Rican student body (60/80 students). Currently, the school also enrolls students of Mexican, African American, and multiple Latino ethnicities from grades 9 to 12 and serves as a "city wide" alternative high school.

Until January of 2003, the high school was located on the second floor of a two-story building purchased by the Puerto Rican Cultural Center (PRCC), an umbrella organization under which various community-based Puerto Rican programs are operated (Ramos-Zayas, 1998). These programs include the high school, an HIV/AIDS awareness project called VIDA/SIDA, and the Division Street Business Development Association (DSBDA) whose role is to encourage Puerto Ricans to relocate and operate their businesses on Division Street—the symbolic home of Puerto Rican Chicago. Since 1974, the PRCC's building was nestled in a predominantly Puerto Rican neighborhood comprised of modest homes and small factories. However, in recent years many of these residents were forced to find cheaper housing in other areas because of gentrification facilitated by developers who purchased many of the former factories and converted them into expensive loft apartments (refer to Alicea, 2001; Flores-González, 2001; and Ramos-Zayas, 2001, for a more complete analysis of gentrification in Puerto Rican Chicago). These same struggles around the effects of gentrification and its implications for schooling and the broader political economy have also recently begun to affect El Puente.

During the late 1990s, as part of a broader strategy to create a distinctively Puerto Rican business center in the Humboldt Park neighborhood, the PRCC's leadership decided to sell the high school building to developers in order to move the high school to *Paseo Boricua*. As a result, the PACHS is now located in a newly renovated building which, unlike the former building, no longer displays the political prisoner murals or nationalist slogans. On the contrary, the new façade is one of beige-textured cement blocks and greatly resembles many of the newly renovated buildings within its immediate vicinity. More importantly, although the new school is smaller than its predecessor, it now has a modern science classroom with new experiment pods and an updated computer lab in addition to six classrooms.

Critical and Caring Curricula

El Puente's approach to developing formal curriculum values and incorporates students' cultural capital or funds of knowledge—what Moll and Greenberg (1990) consider "an operations manual of essential information and strategies households need to maintain their well being" (p. 323). The Sankofa curriculum is a ninth- and tenth-grade English, global studies and fine arts curriculum and is organized around the essential questions, "Who am I?" and "Who are we?" Students explore poetry, art, and cultural histories that address personal identity and the diasporic history of communities of color. They present individual portfolios of art projects, writings, and research about themselves and their family histories.

Additionally, using the arts as a key medium, El Puente organizes annual integrated curricular projects across disciplines and seeks to link them to students' cultural and historical journeys as well as the history and geopolitics of the local Williamsburg community. The Sugar Project, for example, was inspired by a local Williamsburg landmark—the Domino Sugar factory—and linked English, global studies, biology, dance, and visual arts to an exploration of the historical and commercial connections between the Caribbean and Brooklyn:

> Young people studied the history of sugar and its effects (i.e. slavery dependent cultivation in the Caribbean and Latin America) as well as the patterns of consumption in the United States. Students in biology conducted a school-wide survey of the amount of sugar and sugar-based products consumed daily by young people in Williamsburg. The English and Global Studies classes investigated the histories of people who worked on sugar plantations and studied the cultures of resistance which grew out of their struggles. Video, dance and visual arts classes studied the cultural and spiritual expressions that emerged from struggles and oppression related to sugar in Africa, Latin America and the Caribbean. (Pedraza et al., 2001, p. 18)

The culmination of the Sugar project was the Sweet Freedom Sugar Feast, a performance and parade with student stilt walkers, Afro-Caribbean dance, spiritual songs, a skit, and a presentation of a short video produced by young people. This community celebration took place outside in El Puente's community garden, *Espíritu Tierra*, with student-created murals as the backdrop. It combined elements of fantasy, political satire, and traditional culture that together told the history of the people who suffered oppression and resisted in and beyond the sugar fields. Sweet Freedom was an example of the experiential learning that takes place when academic subjects and the arts are integrated (Rivera and Pedraza, 2000, p. 19). In other years, integrated curriculum projects have been organized around themes of the garment industry, biodiversity, media literacy, and power/self-determination.

El Puente's students described the ways in which the Academy's curriculum and pedagogy is relevant to their lives and provide them with important historical knowledge grounded in their identities. Carmen, a Puerto Rican ninth grader, describes the significance of the "who am I book," an element of the Sankofa curriculum and how she was engaged through lessons on the Taíno indigenous people of the Caribbean:

> We are exploring ourselves. In global studies, we're learning about the Taíno—the indigenous people of Cuba, Dominican Republic and Puerto Rico. We're learning about our roots. My facilitator brought in some Taíno artifacts and showed us all the weaponry and stuff, and it was cool. El Puente finds a way to teach you, and you have fun at the same time. In all the classes, we are learning about who we are because we write this book, the "who am I book." That's stuff I never really thought about before.

Similarly, the PACHS curriculum places emphasis on students being able to analyze and transform their lives, the lives of others, and the communities in which they reside from critical perspectives through the lenses of racial/ethnic, cultural, and politically nationalist affirmation. The curricular objective of decolonizing the mentality and actions of students and community members is formally conceptualized into three components named "Identity," "Cognitive Skills," and "Action."

The "Identity Component" of the curriculum stresses the importance of students analyzing their social realities as Puerto Ricans, Mexicans, African Americans, or students who may identify themselves as being of multiple Latino ethnicities. Student-based social analysis occurs through the offering of Puerto Rican, Mexican, and African American history and literature courses. Teachers utilize multiple texts that present topics through an alternative lens such as the work of Howard Zinn (*A People's History of the United States*) in addition to standard history textbooks. Other teachers use texts that specifically address African American and Mexican historical issues.

The "Cognitive Skills" component of the high school's curriculum reflects a more traditional public high school curriculum and includes biology, chemistry, algebra, basic arithmetic, geometry, calculus, and trigonometry, among other courses. The third and final component of the curriculum is called "Action" and is implemented through classes that encourage hands-on student experiences such as photography, art, journalism, and video production. Other activities within this component involve student participation in community events such as community protest marches, community cleanups, and cultural events. Although students are not obligated to participate in these community events, many of the students who did were praised and given extra credit toward a higher grade in their "Unity for Social Analysis" class. This class stresses the importance of connecting project-based learning to community involvement such as beautification projects and civil disobedience regarding the release of U.S. political prisoners and gentrification. Melissa, a Puerto Rican PACHS senior, commented that her previous high school experiences were culturally irrelevant and that more high schools should include courses that specifically address the subjugated sociohistorical and political realities of their students.

> At my old public high school I had no idea who Pedro Albizu Campos was or who Lolita Lebrón was. I had no idea who these people were. Somebody came up to me and asked if I knew what the *Grito de Lares* was. I was like, "What is that?" My Puerto Ricaness was challenged when they asked me, "You're Puerto Rican, right?" They then told me I should know this stuff. None of this was ever taught to me. I think public schools should have different kinds of history classes like African history, South American history, and other stuff that isn't normally taught.

Damien, a tenth grader, also observed that his previous public high schools did not undertake any serious attempts to weave culturally relevant curricula in courses. In fact, he felt that these schools were, in essence, "brainwashing" students like him to accept a set of mainstream realities that ran counter to ones he wanted to learn more about.

> In Hartford and Chicago I was brainwashed. There was always a side of me that wanted to learn more about my culture. I wanted to learn more than what the schools were telling me. It was at the PACHS that I heard of Puerto Rican writers like Lola Rodríguez de Tió, Luis Muñoz Marín, and Dr. Pedro Albizu Campos. But when I was in school they never taught me what I wanted to know. They would only teach me to pledge allegiance to the United States flag and sing the Star Spangled Banner. These are all lies. They never told me about the splendid little war and how the United States went into Cuba, Puerto Rico, Guam, and the Philippines. They never told me how they went in and took Hawaii. In the public schools they never taught me about the slaughtering of people in Vieques. The teachers always tried to make the United States seem all high and mighty.

Although Melissa and Damien expressed their desire that schools radically transform their curricula in ways that would address alternative epistemologies, PACHS and El Puente students also felt it was important for teachers to not only have a passion for the provision of academic content but to also build and sustain high-quality interpersonal relationships with them. This powerful combination of high academic expectations and meaningful student-teacher interpersonal relationships form the basis of authentic caring as an alternative to the traditional schooling that many students of color in urban schools experiences on a daily basis.

Authentic Caring as Explicit Curriculum

Time and time again, student informants articulated the importance of authentically caring relationships with their teachers/facilitators and described their relationships with teachers at PACHS and El Puente in contrast to their experiences of non-caring in prior schools. These experiences constitute what we call an explicit (or not-so-hidden) curriculum that counteract the informal and formal practices that historically marginalize Latino/a students. Bowles and Gintis (1976) describe the hidden curriculum as the nature of social relations in classrooms and schools that transmit messages legitimizing class-based positionalities in regard to work, rules, authority, and values that maintain capitalist sensibilities. The hidden curriculum, then, becomes the mechanism by which students learn their place in the economy, accept their position, and develop the necessary skills for their role in the labor force.

Alternatively, student voices reveal that PACHS and El Puente's formal and informal curricula are organized in ways that encourage students' active engagement as members of a learning community. Emphasis on instrumental relationships (Stanton-Salazar, 2001) and the Latina/o cultural value of *personalismo* (Santiago, Arredondo, and Gallardo-Cooper, 2002) create conditions that student informants describe as transformative[3] relative to their prior experiences in traditional schools and their knowledge of schools that friends and relatives attend.

For example, Melissa, a Puerto Rican PACHS senior, described the ways in which teachers and students had marginalized her in a school characterized by deep class divisions:

> The year I left the other school, they had taken too many students. Most of my teachers cared about the richer and better students. The ones who were poor or at the bottom were ignored. The teachers didn't care because they put down students and called them names. One time a teacher said that I would become nothing but a future statistic—pregnant or raped somewhere. The White students heard that and started calling me "stat." I was also the only Puerto Rican in that advanced science class. Things were really bad. I had to get out.

Kathy, a multiethnic Latina (Mexican and Puerto Rican) who graduated from the PACHS in 1985, also described how her experiences with non-caring and non-Latino teachers facilitated her exit from a traditional high school and subsequent enrollment at PACHS:

> The teachers in my other high school were mean. They would speak down to you. I had no Latino teachers. My teachers didn't even know my name. If they wanted to get my attention, they would poke at me or yell at me. After a month of this shit, I was like, "I'm outta here!"

These reported experiences resonate with other researchers' (Katz, 1999) observations regarding teachers who had no knowledge or interest in becoming familiar with their students' social and cultural realities and, as Kathy asserted, this alienation was reinforced by a lack of any Latina/o teacher presence.

Teresa, an African American El Puente senior, linked a critique of her previous school to her positive experience at El Puente and its articulation of a mission "to inspire and nurture leadership for peace and justice." She provides an analysis of the tacit mission of her former school.

> This whole thing of basically having a mission is different right there—you can't ask no other school like "what's your mission" 'cause I don't really think they (my former school) have none. I think it's just to get those students who aren't doing well out—'cause they also push a lot of students ahead without them making their grades. And I haven't seen that done in this school.

Teresa's critique of the lack of a mission of her former school and the practice of "pushing students ahead" suggests that educational engagement at El Puente is related to both high expectations and a high level of support placed on her by facilitators. For Melissa, Kathy, and Teresa, teacher apathy and low expectations contributed to academic alienation and their eventual exit from traditional high schools. As Teresa argues, this sentiment is reflective of a "push out" rather than "drop out" experience and yet these student observations suggest that something about the social

> organization and pedagogy of these two schools generate a culture of student "drop-in"[4] and academic engagement. Other students suggested that the substance of this engagement in school emerges from high quality instrumental relationships with teachers. (Stanton-Salazar, 2001)

School as *Familia y Communidad*: The Importance of Student-Teacher Relationships

Student informants from both schools described experiences of educational caring linked to high academic expectations and mentorship as central features. These reported experiences highlight an emphasis on high-quality interpersonal relationships at the two schools. This emphasis, we argue, emerges from the Latina/o cultural value of *personalismo* that Santiago, Arredondo, and Gallardo-Cooper (2002) describe as having important institutional implications.

> High importance is given to the qualities of positive interpersonal and social skills that family members, both nuclear and extended, maintain mutual dependency and closeness over a lifetime. The valuing of warm, friendly and personal relationships has important implications for how Latinos respond to environments (e.g. hospitals, mental health agencies, etc.) that are quite often impersonal and formal. (p. 44)

By institutionalizing an ethic of *personalismo*, PACHS and El Puente staff transcend the boundaries of traditional schooling and create social conditions and relationships that are more aligned with students' cultural orientations and that overlap with extended family life. Additionally, because these schools emerged out of community struggles for denied educational rights, they embody important social and cultural protective features. Reflecting these struggles the terms *respect, friendship*, and *family* frequently and compellingly arise in the interviews and conversations we had with students. Students often described their relationships with facilitators as *like a friend*, *like family*, or *like a parent*. Ricardo, for example, a Dominican-born sophomore, expressed that El Puente was a caring school because of the sense of family and community that his teachers fomented. Suggesting that traditional power relations between teachers and students are also subverted at El Puente, he used a "parent-son" metaphor to illustrate his experiences:

> El Puente's a very good school because the teachers really treat you like a family. In some other schools you gotta call the teacher Mr. Rodríguez or Mr. This, Ms. That. At El Puente, you call your teacher by their first name, like one of your friends. If you got a problem [in other schools] they tell you, "You can do anything you want, after my class." At El Puente they don't do that. If you do something bad, they all sit with you and have a meeting with the principal and they try to help you in whatever you need. They sit with you and talk to you like it was a parent to a son.

Because he feels cared for by his facilitators Ricardo does not see school as a place where adults focus narrowly on academic content and routines or where he must be guarded and distrustful of authority figures that will punish or suspend him. Moreover, by abandoning the use of formal surnames, El Puente's facilitators communicated their interest in redefining the traditional "power over" model of student/teacher relations for a "power with" model (Kreisberg, 1992).

Kathy, a 1985 PACHS graduate, also considered her teachers to be caring because they were willing to be learners with their students and because there did not exist a hierarchical power division between student and teacher.

> The teachers don't have that aura of being superior because they belong to the faculty or administration. For me, the teachers acted like co-students. They cared because they were there to work with you and learn with you. It was a different feeling than what I got at the large public school I attended.

Kathy's statement reflects the PACHS commitment to utilize an educational philosophy derived from the work of Paulo Freire known as critical pedagogy, which emphasizes the teacher's dual role as facilitator and learner (Freire, 1970). Similarly, El Puente refers to its teachers as *facilitators* because they facilitate learning rather than utilize banking methods and "fill empty vessels." Students also suggest that such views about the role of teachers have important implications for creating a learning community where students also support each other's learning. Pura, another 1985 PACHS graduate, reinforced how the PACHS facilitated this sense of community among students:

> The students here looked out for each other and we worked for one another. Everything was done together. If a decision or situation had to be made or

resolved then there was a discussion in the school and the decision was made together. I still remember having special school events together. We looked at each other as being part of a family.

Similarly, Carmen, a freshman at El Puente, commented:

> El Puente's about loving and caring, support, community, like we're all one and united. It's the way people interact with each other, you know? The facilitators are good. They care about the students. Basically, they treat you like friends. You can call them by their first name, just like they call you by your first name. It mainly has to do with respect. They're caring.

Critics of such a strong emphasis on interpersonal relations may express concern that such highly personalized and informal relationships, like the ones valued at both schools, might have then potential to diminish boundaries and authority relationships between youth and adults. Students, however, reported that facilitators at El Puente negotiate relationships that are indeed bounded, respectful, and evocative of student development. Teresa observed:

> I think they come down to our level in a mature way. Like they can hang out with us and talk to us on our same level- but it's like they're not really with us. They know how to have a good time with us—how to talk to us- how to find out what we're thinking but at the same time not really act childish. They still know their place—have a good time and let the student know that they are older and they do have a certain respect- so if you're sitting down with a facilitator you don't cuss or anything.

The notion of school as "familia y comminidad" is facilitated at both schools by an explicit commitment to engaging Latina/o students in learning through close, high-quality interpersonal relationships between teachers and students. Trina, an African American El Puente freshman, commented on these three important conditions:

> It's nice, it's different, and very unique. It's a loving school. When you come into this school everybody accepts you. Everybody kind of embraces you and takes you in. It's a lot smaller than the average high school. And I think we are closer and we get along more because everybody gets to know each other. Everybody is familiar with the teachers and the staff. They [facilitators] take their time with you and they ask you if anything is bothering you. They're caring.

Trina's experience of acceptance and her observation that "we get along more because everybody gets to know each other" suggests that interactions among students are shaped not just by the size of the school but also by the nature of relationships between facilitators and students. Both PACHS and

El Puente students described powerful experiences of being cared for by teachers at their respective schools.

Students explained that caring teachers offered them guidance and friendship inside and outside the classroom, held them to high academic expectations, and demonstrated a sense of solidarity by being active co-learners and facilitators rather than authoritarian teachers. These observations reveal a strong emphasis on *personalismo* as an informal curricular practice and strongly suggest that the cultural orientations and values at the two schools are closely aligned with the expectations that Latina/o families have of schools—an emphasis on social relations, self-awareness, and respect in addition to academic preparation. Valenzuela (1999) articulates these expectations by using the Spanish term *educación*:

> Educación is a conceptually broader term than its English language cognate. It refers to the family's role of inculcating in children a sense of moral, social and personal responsibility and serves as the foundation for all other learning. Though inclusive of formal academic training, educación additionally refers to competence in the social world, wherein one respects the dignity and individuality of others. (p. 23)

Based on Valenzuela's definition, the notion of *educación* or *ser bien educado/a*[5] (to be a well-educated person) is deeply rooted in relational and social ties characterized by respect (respeto) and confianza (mutual trust) (Moll and Greenberg, 1990), which student informants revealed are operationalized at PACHS and El Puente and facilitate their academic engagement and achievement.

Conclusions

The voices of PACHS and El Puente students presented in this chapter reveal that authentic forms of caring based on Latina/o values and struggles for educational rights are embedded in the formal and informal structures and curricula of both schools. Students consistently reported that they were significantly engaged in the learning process in these two schools through quality interpersonal relationships with adult teachers and facilitators and that these relationships were characterized by high academic expectations of the students by staff.

This emphasis on *personalismo* and high expectations is consistent with our articulation of the term "hard caring" or a form of *critical care*, because motivation finds origins in a critique of traditional forms of schooling in the lives of marginalized Latina/o youth. Moreover, these practices result in explicit, rather than hidden, commitments to creating curriculum that affirms student identities as well as culturally relevant pedagogy (Ladson-Billings, 1995). Moreover, these particular schools are very different than their small, progressive, White-led high schools because they are actively led by leaders of color who are aware of their students' daily lives/struggles as a result of residing within the very neighborhoods where the schools and their students are situated. Thus, these findings are important because they strongly support our argument that while schooling on a smaller scale is an important condition for engaging Latina/o youth in learning, what goes on within those small structures is infinitely more important. In this regard, we believe that our analysis of student and critically caring practices at El Puente and PACHS contribute to McKamey's notion of a process theory of caring by advancing the concept of critical care as one that is motivated by community protective interests and considers cultural and community contexts where these are too often ignored. Moreover, through a notion of hard caring vs. soft caring, we believe it useful in both critiquing emotionally laden origins of teacher caring as opposed to relevant forms of caring for marginalized students that are characterized by supportive interpersonal relationships and high academic expectations.

Additionally, we feel it is of the utmost importance that communities of color have a primary role in the creation of small high schools because they have firsthand knowledge of their students and their sociopedagogical and political interests. Because PACHS and El Puente were organically created and sustained for and by community members, as opposed to large, impersonal, and bureaucratic school districts, students and teachers were able to authentically privilege and honor their respective funds of knowledge and dismantle the subtractive schooling (Valenzuela, 1999) practices that were so commonplace in their previous comprehensive high schooling experiences. Thus, we strongly encourage small school reform leaders to steer their pedagogical and political efforts toward communities of color and away from the large comprehensive high schools that have consistently failed them. Therefore, we unequivocally call for the death of comprehensive high schools and call to life small schools that have the potential to serve urban youth of color and the communities they call home.

Notes

* We wish to thank Daniel Liou, of the University of California-Los Angeles, for his helpful insights and suggestions regarding earlier drafts of this chapter.

1. Popularized by Nuyorican Poet Mariposa, the term DiaspoRican connotes the increasingly disperse and evolving nature of Puerto Rican identity within the United States and abroad (Valldejuli and Flores, 2000).
2. For example a *New York Daily News* reporter observed, "through years of riots, shootings, stabbings, parent protests and facility crises, Eastern District became synonymous with the wilder aspects of the decline of urban education" (Williams, 1998).

3. By *transformative*, we mean the extent to which students perceive that their educational experiences (not necessarily outcomes) are better than their experiences would be elsewhere.
4. This term is adapted from Fry (2002) who, from his analysis of Latino dropout data, argues that a substantial number of Latino immigrant students never dropped out of school because they never had the intention of attending school (dropping in) when they arrived in the United States, rather they sought workforce participation.
5. In Latin America (as within U.S.-based Latino/a communities), use of the term *educación* and "*ser educado/a*" can relate to social class and race-based differences and strongly implies racial inferiority toward persons of African and indigenous decent. Notwithstanding this limitation, I argue that, within the U.S. schooling context, Valenzuela's usage of the term provides a useful distinction in understanding Latino/a communities' orientations toward education and expectations from public schools.

References

Alicea, M. (2001). *Cuando nosotros vivíamos . . .* : Stories of displacement and settlement in Puerto Rican Chicago. *Journal of the Center for Puerto Rican Studies* 13 (2): 166–95.
Abi-Nader, J. (1990). "A house for my mother": Motivating high school students. *Anthropology and Education Quarterly* 21 (1): 41–58.
Ayers, W., Klonsky, M. and Lyon, G. (2000). *A simple justice: The challenge of small schools.* New York: Teachers College Press.
Ayers, W., Hunt, J.A., and Quinn, T. (1998). *Teaching for social justice.* New York: The New Press.
Fine, M. (2000). A small price to pay for justice. In W. Ayers, M. Klonsky, and G. Lyon (eds.), *A simple justice: The challenge of small schools* (pp. 168–79). New York: Teachers College Press.
Flores-González, N. (2001). *Paseo Boricua*: Claiming a Puerto Rican space in Chicago. *Journal of the Center for Puerto Rican Studies* 13 (2): 6–23.
——— (2002). *School kids/street kids: Identity development in Latino students.* New York: Teachers College Press.
Freire, P. (1970). *Pedagogy of the oppressed.* New York: Continuum.
Katz, S. (1999). "Teaching in tensions:" Latino immigrant youth, their teachers, and the structures of schooling. *Teachers College Record* 100 (4): 809–40.
Kreisberg, S. (1992). *Transforming power: Domination, empowerment, and education.* New York: SUNY Press.
Ladson-Billings, G. 1995. Toward a theory of culturally relevant pedagogy. *American Educational Research Journal* 32 (3): 465–91.
McKamey, C. (2002). *Competing theories of care in education: A critical review and analysis of the literature.* Unpublished manuscript, Harvard University.
Mohr, N. (2000). Small schools are not miniature large schools. In W. Ayers, M. Klonsky, and G. Lyon (eds.), *A simple justice: The challenge of small schools* (pp. 168–79). New York: Teachers College Press.

Moll, L.C. (1990). *Vygotsky and education: Instructional implications and applications of sociohistorical psychology*. New York: Cambridge University Press.
Moll, L.C. and Greenberg, J.B. (1990). Creating zones of possibilities: Combining social contexts for instruction. In L.C. Moll (ed.), *Vygotsky and education: Instructional implications and applications of sociohistorical psychology* (pp. 319–48). New York: Cambridge University Press.
Moll, L.C. et al. (1992). Funds of knowledge for teaching: Using a qualitative approach to connect homes and classrooms. *Theory into Practice* 31 (1): 132–41.
Nieto, S. (1998). Fact and fiction: Stories of Puerto Ricans in U.S. schools. *Harvard Educational Review* 68 (2): 133–63.
——— (2000). Puerto Rican students in U.S. schools: A brief history. In S. Nieto (ed.), *Puerto Rican students in U.S. schools* (pp. 5–38). Mahwah, NJ: Lawrence Erlbaum.
Noddings, N. (1984). *Caring: A feminine approach to ethics and moral education*. Berkeley: University of California Press.
——— (1992). *The challenge to care in schools: An alternative approach to education*. New York: Teachers College Press.
Pedro Albizu Campos High School Teacher Policy Manual (no date). Chicago.
Ramos-Zayas, A.Y. (1998). Nationalist ideologies, neighborhood-based activism, and educational spaces in Puerto Rican Chicago. *Harvard Educational Review* 68 (2): 164–92.
Rivera, M., and Pedraza, P. (2000). The spirit of transformation: An education reform movement in a New York City Latino/a community. In S. Nieto (ed.), *Puerto Rican students in U.S. schools* (pp. 223–46). Mahwah, NJ: Lawrence Erlbaum.
Rodríguez, G., Antrop-González, R., and Reyes, A. (2006). Documenting Latino achievement: Recent research on Puerto Rican and Mexican-origin high school students. In F. English (ed.), *Sage encyclopedia of educational leadership and administration*. New York: Sage.
Rolón, C. (2000). Puerto Rican female narratives about self, school, and success. In S. Nieto (ed.), *Puerto Rican students in U.S. schools* (pp. 141–66). Mahwah, NJ: Lawrence Erlbaum.
Santiago-Rivera, A., Arredondo, P., and M.Gallardo-Cooper (2002). *Counseling Latinos and La Familia: A practical guide*. Thousand Oaks, CA: Sage.
Stanton-Salazar, R. (2001). *Manufacturing hope and despair: The school and kinship networks of U.S.-Mexican youth*. New York: Teachers College Press.
Thompson, A. (1998). Not the color purple: Black feminist lessons for educational caring. *Harvard Educational Review* 68 (4): 522–54.
Valenzuela, A. (1999). *Subtractive schooling: U.S.-Mexican youth and the politics of caring*. Albany: SUNY Press.
Valldejuli, J.M., and Flores, J. (2000). New Rican Voices: Un Muestrario/A Sampler at the Millenium. *Journal of the Center for Puerto Rican Studies* 12 (1): 49–96.
Williams, L. (1998). Last graduation ends long, troubled legacy. *New York Daily News*, June 29, 1998: Suburban page 1.

Chapter 6

Soul Making in the Comprehensive High School: The Legacies of Frederick Wiseman's *High School* and *High School II*

José R. Rosario

Introduction

What should we make of ourselves? This question is relevant not only to the growth of nations, as Richard M. Merelman (1984) suggests in writing about America. It is equally relevant, and perhaps more importantly so, to the growth of individuals. In *The Ethics of Identity* (2005), Kwame Anthony Appiah describes the matter in the following way:

> Each of us has one life to live; and although there are many moral constraints on how we live our lives . . . these constraints do not determine which particular life we must live. . . . This means at least two things. First, the measure of my life, the standard by which it is to be assessed as more or less successful, depends . . . on my life's aims as specified by me. Second, my life's shape is up to me (provided that I have done my duty toward others), even if I make a life that I could have made . . . Thoughtful friends, benevolent sages, anxious relatives will rightly offer us both assistance and advice as to how to proceed. But it will *be* advice, not coercion, that they justly offer. And just as coercion will be wrong in these private circumstances, it will be wrong

when it is undertaken by governments interested in the perfection of their citizens. (p. xii)

In Appiah's ethics of individuality, we all share a project in "soul making," in constructing an identity or creating a life for ourselves. This project is ethical as opposed to moral in that the soul making it entails "has to do with what kind of life it is good for us to lead" and not about "what we owe to others" (p. 230). For Appiah, who is following Dworkin (2000) in this regard, the former is about "convictions," the latter about "principles." How we approach the project as an ethical enterprise is ultimately an existential or individual concern that others can support but not determine or control. We create a life for ourselves, according to Appiah, by working with and working on what we have been given.

Among the institutions that play a role in this process, one that is especially important, is the school. It is this agency, and particularly the comprehensive high school, which mediates the transition of our nation's youth from adolescents to the beginnings of adulthood, that will be the focus of this chapter. How has the comprehensive high school figured in this ethical project we all face? What is its legacy? What has it given our young? There are as many answers to these questions as there are perspectives. How we answer depends largely on what we take schools to be about and on what we believe they have accomplished. In *High School* (1968) and *High School II* (1992), Frederic Wiseman, perhaps the most celebrated of American documentary filmmakers practicing the *cinéma vérité* style, provides us with visual portraits that pose answers to these queries. In this chapter, I deconstruct and explore these portraits in an effort to understand how the comprehensive high school has figured in our young's ethical struggle to make something of themselves. The focus is not on the aesthetics of *High School* and *High School II* as films or works of art. My interest is limited, rather, to treating the films as testaments, visual documents, or chronicles of school life rooted in two very different settings and historical contexts. I aim to uncover what Wiseman himself referred to as the "theories" behind his films (Levin, 1971; Grant, 1992; Benson and Anderson, 2002), and end with an illustrative narrative of the changing dynamics in American secondary education.

High School and *High School II* point to different political rationalities of how a high school should proceed in fabricating character, what I am also calling, following Appiah (2005), as well as Popkewitz (1998), the "soul." To explore these politics requires that I venture into the analytics of governmentality, what Foucault (1991) called "the conduct of conduct." From the perspective of governmentality, argues Rose (1999), governance defines any practice that aims "to shape, guide, direct the conduct of others, whether these be the crew of a ship, the members of a household, the employees of a boss, the children of a family or the inhabitants of a territory. And it also embraces," he proceeds to add, "the ways in which one might be urged to and educated to bridle one's own passions, to control one's own instincts, to govern oneself" (p. 3). My shorthand for the term is *soul making*.

In schools, soul making operates through two different systems of structuring. Grant (1988) speaks about these systems in terms of intellectual and moral ordering. The school's intellectual order consists of "the skills, concepts, and knowledge that are taught or imparted to the students; the moral order, to the impact the school has on their conduct, character, and moral beliefs" (p. 180). In my analysis of *High School* and *High School II*, the focus is on the school's moral order, in uncovering how it gets constructed and functions to govern and fabricate conduct. I draw attention to scenes that embody what Jackson (1988) calls "sermonettes" (Jackson, 1988), which I interpret as consisting of vignettes illustrating forms of discourse that inscribe the ethical narratives students are expected to use in fashioning their identities. In the case of *High School*, I argue, the sermonettes appear as part of a governmentality that operates through policing. Sermonettes are delivered to exert greater power and control over ethical constructions. In *High School II*, on the other hand, the objective of sermonettes is seemingly different: rather than regulate soul making by policing, they aim to control it by community, a variation on what Rose (1999) calls "government through community" (p. 176). The films are deliberately silent on which of these methods contribute more to molding the ethical projects of our youth, and so am I—for I aim to give voice to a less ambiguous point: that the films signify a clear shift in how schools ought to govern the fabrication of student identity—from regulation through policing to regulation through community.

The Films

High School and *High School II* are strikingly different films about strikingly different high schools. Filmed and released in 1968 at the height of the civil rights movement and the Viet Nam war, *High School* is the shorter of the two (75 minutes and 43 scenes), and the second of Wiseman's 34 films. *High School II* is considerably longer (220 minutes), was filmed in 1992, and released in 1994. *High School* was filmed at Northeast High School, a large, comprehensive, predominantly white, middle-class high school, in Philadelphia, Pennsylvania. *High School II*, on the other hand, was filmed in inner-city New York at Central Park East Secondary School (CPESS), a considerably smaller and predominantly minority institution. Portions of *High School II* were filmed during the Los Angeles riots that resulted from the verdict in the Rodney King trial, and, although longer, the film consists of only one more scene than *High School* (44 versus 43). In summarizing the key differences between Northeast High and CPESS, Wiseman put it this way:

> One difference is size—North East had approximately 4,000 students; Central Park East has about 480 students. The principal difference, though, is ideology. Central Park East does have a carefully thought-out theory of education which is related to all aspects of life at the school. It's summarized in the "habits of mind," posted in every room. Their educational philosophy includes such issues as class size, the content of curriculum, personal relationships between faculty and students, and the relationship between parents and the school. My impression is that all aspects of the school are not only thought through, but constantly reconsidered and reevaluated in the light of experience and theory. There is an enormous sensitivity to the realities of the students' life-situations. (Lucia, 1994, p. 6)

Both films were intended to portray "good" schools (Handelman, 1970; Levin, 1971; Rosenthal, 1971; Lucia, 1994; Krupnick, 1995; Benson and Anderson, 2002). In the case of *High School*, for example, Wiseman sought to film "the kind of school that the blacks were aspiring to rather than a ghetto school" (Handelman, 1970, p. 6). He "didn't want to do a black school, because that scene had been overworked," and he "felt it would be more interesting instead to have a look at what was thought to be a good school, and to see the kinds of values being taught to the students" (Rosenthal, 1971, p. 70). He settled on Northeast High because "by common consensus" the school "was thought to be one of the two best high schools in Philadelphia" (Rosenthal, 1971, p. 70). For *High School II*, Wiseman wanted a school "in a minority community." However, he still desired "a high school that was succeeding" in order to avoid "all the cliché problems associated with urban areas" (Krupnick, 1995, p. 5).

High School and *High School II* are emblematic of two stages in the evolution of America's high schools in the last half of the twentieth century. In the former, the viewer finds a large, primarily white, authoritarian, and comprehensive school; in the latter, a small, primarily minority, democratic, and much less comprehensive school. Both seem to stand as bookends to the tumultuous changes the American high school experienced between the years marking the two films. As Cusick (1983), Grant (1988), and Rosario (2000) have shown, the American shifts in education policy to engineer a more egalitarian society during the 26 years separating the two films did much to deconstruct the notions of community evident in *High School* and pave the way for the more progressive vision of secondary schooling captured in *High School II*. But as portrayed in the two films, the schools also signify different truth regimes about how to govern the young and shape their identities. And it is to this theme, beginning with *High School*, that I now turn.

High School: On Soul Making by Policing

I am not alone in sharing the conventional wisdom about *High School*: that it is a film about "institutional power" (Benson and Anderson, 2002) and "depersonalization" and "ideological indoctrination" (Grant, 1992). From the film's opening shots to its last, Northeast High School is rendered as a militaristic "factory" aiming to control the identities students carve out and construct for themselves with the social capital they are given. According to Benson and Anderson, the school's militarism is exercised by positioning students into a "double bind": "students receive inconsistent directions in a situation in which they must try to obey the directions, but in which they cannot call attention to or even conspicuously notice what is happening" (pp. 138–39). The strategic end of this regulatory practice is to mold and shape character, "to build a system of control that will last the students for a lifetime" (p. 138). Power is thus exercised through a process of hegemonic "tyrannies" consisting of "direct orders, the imposition of a routine, humiliating reprimands, absurdly manipulated language, mechanical teaching that reduces learning to a stultifying process of repetition, and, most hurtfully, through a constant confusion of thought and feeling about sex and gender" (p. 139). As a result, students become so saturated with the school's ideology of control that they are unable to problematize the institution's governing practices and contest them. Their only defense, argue Benson and Anderson, "is to disconnect, to resort to daydreams, narcissism, and apathy" (p. 139). Weisman himself was to characterize this disaffection as "the terrible passivity of the students." He expressed his disbelief in what he found as follows:

> One of the most depressing things about the whole experience there was the total passivity of the students. It wasn't something I selected. It was around the whole place and I couldn't believe it. This was in the spring of 1968, just around the time that Johnson had withdrawn and the bombing of North Vietnam had been halted. McCarthy had won the New Hampshire primary. Bobby Kennedy was going in. King had been killed. There was a fantastic amount going on in the country of great importance. There was a lot going on in Philadelphia, but these kids were really out of it. The only lively bunch that I found were those kids in the scene toward the end of the film. That was in extra-curricular, in what they call "the human relations club" that met after school. Those kids were easily the liveliest bunch, and they were all just about drop outs. (Rosenthal, 1971, p. 75)

Earlier, I wrote of governmentality as any practice that seeks to regulate and script conduct. I also characterized such practices as engaging in soul making. In characterizing what I and others see at Northeast High, I suggest that governmentality at the school amounts to *soul making by policing*. In saying this, it is important to note that this is a judgment of the school's practice and not its theory. It is important not to lose sight of Wiseman's observation about the gap he found in the school between ideology and practice: "In one way the film is organized around the contrast between the formal values of openness, trust, sensitivity, democracy, and understanding, and the actual practice of the school which is quite authoritarian" (Rosenthal, 1971, p. 71). The conventional wisdom on *High School* is curiously silent on this contradiction, and I am not prepared to provide an answer here. My concern for now is unpacking the governing orientation Wiseman has uncovered in his "voyage of discovery" (Grant, 1992).

I believe the school's authoritarianism is traceable to the premise on which it rides: that policing is a legitimate and just means to the construction of identities. School personnel seem to derive legitimacy for the claim from the political Right they take as logically and automatically flowing from their positions as adults endowed with the authority to exercise power without fear of retribution. This paternalism is evident throughout the film, but it is particularly salient in three of the film's most celebrated scenes: 6, 9, and 10.

In each of the three scenes, a school official is positioned to answer the exercise of power by other adults in the school, and the response in each case is legitimation of the action by in turn repositioning and constructing the other as lacking the moral standing to dissent. In one such scene, for example, a student tries to explain why he cannot take a gym class. The administrator ignores his appeals, silences him, and orders him to class on the grounds that "we'll determine whether you take exercise or not." Although he assures the student "no one's going to put" him "in an uncompromising position," he expects him to "come dressed in a gym outfit." When the student answers he "would," the administrator suspends him. He had been cast as having no voice in school matters, and had been commanded "not to talk and just listen." Saying "I said I would" was sufficient for the administrator to judge and construe the student as an insubordinate worthy of punishment. And he appears to act in that fashion only by virtue of the authority he claims as an adult and administrator in the school (Grant, 1992, pp. 51–54).

In another scene, a student who talked back to a teacher is seen before the same administrator. He is challenging the teacher's punishment on the grounds that he did nothing wrong. The administrator tries to convince him to show "some character," accept the punishment, and demonstrate he "can be a man" who "can take orders." He had, after all, shown "poor judgment." "When you're being addressed by someone older than you are in a seat of authority," he sermons to the boy, "it's your job to respect and listen." When the boy objects that "it's all against [his] principles" to take orders and that "[he has] to stand for something," the administrator retorts that his "principles aren't involved here." More at stake was "proving to be a man," conforming to "rules and regulations," not fighting "with a teacher," and requesting "permission to talk." If "the teacher felt [he was] out of order, and in her judgment [you] deserved a detention," he did not "see anything wrong with assigning [him] a detention." The student had only to "prove yourself" and "show that you can take the detention when given it." The teacher had not asked for his "blood" or that he "jump from the Empire State Building." The teacher was merely asking "for a little bit of time, to help [him] out." If there was a perpetrator here, it was he. He is constructed as lacking character, not yet a man, and is convinced to take the detention, which he accepts "under protest" (Grant, 1992, pp. 57–60).

In the last of the three scenes, the viewer finds a father and mother protesting their daughter's failing grade to a school counselor. The father finds it difficult to understand how a teacher can judge his daughter's work as "fabulous" and still fail her. He questions whether it is "compassionate" and "fair" for a teacher to be this contradictory. The counselor ignores the question and reframes the problem by suggesting he may be raising the wrong question. "Why don't you put it in reverse?" he asks. "Why didn't you say, why didn't you say that a student who can write fabulous papers shouldn't flunk? Shouldn't do things that would cause her to flunk?" When the father asks that the counselor concede "that a girl who receives 'Fabulous' and all these marks shouldn't flunk," the counselor refuses. He will admit only "that the teacher in reading these papers thought they were fabulous, but that the total mark involves more than just those papers." He recasts the daughter as underperforming and repositions the father to concede the daughter's failure. He also constructs the father as the person jeopardizing the daughter's "happiness" by imposing on her "preconceived values and dreams." The counselor wants him "to understand that part of [*his*] job is to deal with Rhona as a sympathetic and understanding father but recognizing her limitations and the fact that she's a person in her own right and that if you impose too much your desires on her, even without pushing them or forcing her, she may react in a way which may be damaging to her too." The source of the problematics here, in other words, is not entombed in the act of an uncaring and unsympathetic teacher; it is contained, rather, in the dynamics created by an underachieving daughter and an overly paternalistic father (Grant, 1992, pp. 55–57).

Easily the most celebrated of *High School* scenes is the last, where the viewer finds the school principal before an assembly of the faculty prepared to illustrate Northeast High's capacity to manufacture character. She uses a letter from a former graduate about to land in Vietnam. Here is an excerpt of what she reads:

> ... We are going to be dropped behind the DMZ, the Demilitarized Zone. The reason for telling you this is that all my insurance money will be given for that scholarship I once started but never finished, if I don't make it back ... I have been trying to become a Big Brother in Vietnam ... I only

hope that I am good enough to become one . . . I really pray that the young men in your cooking classes will use this change very well. . . . My personal family usually doesn't understand me. . . . They don't see . . . why I have to do what I do. . . . Am I wrong Dr. Haller? If I do my best all the time and believe in what I do, believe that what I do is right, that is all I can do. Please don't say anything to Mrs. C. . . . She would only worry over me. I am not worth it. I am only a body doing a job. In closing, I thank everyone for what they have done for me. . . . Please forgive my handwriting. I am a little Jumpy . . . (Grant, 2006, pp. 87–88)

For the principal, the student's farewell is a clear sign all is well at Northeast High. She believes all would agree they are all "very successful" at what they do. Based on Wiseman's images, I am inclined to concur with the principal's fabrication as well. It is difficult to refute, given the kind of character policing seen in scenes 6, 9, 10, and a number of others, the school's strong hand in the fabrication of the young man's ethical and moral constructs.

High School II: On Soul Making through Community

There is no doubting the images on the screen: *High School II* is about a radically different school. Gone are the scenes of a discipline office and the tense confrontations between students and disciplinarian. In a film over three hours long, only one scene (36A), about forty-six seconds long, concerns a teacher-student altercation. Given the widely shared disparaging views of inner-city schools, one would expect to see more. Yet, we do not, for Wiseman's lens is on a school that manages deviance differently than was the case at Northeast High.

CPESS is a school of choice: it is open to any student who applies and commits to its principles. The school is also the outcome of two convergent forces that emerged in the years between filming. Both were responses to the aftermath left by the egalitarian push for fairness and equity in public education, which some analysts saw as a deconstruction of the old orders of schooling (Cusick, 1983; Grant, 1988; Rosario, 2000).

One was the "school as community" movement that was traceable to the work of John Dewey and the rise of communitarianism in the 1960s (Merz and Furman, 1997; Rosario, 2000). For the proponents of this coalition, schools were highly bureaucratized and depersonalized places that alienated and demeaned students. To be effective, schools needed to (1) acknowledge the importance of meaning and value in human development, and (2) be organized around communal principles. Good schools were schools that had a "strong positive ethos," a "sharing of attitudes, values, and beliefs that bond disparate individuals into a community" (Grant, 1988, p. 117).

The second force, the "small school movement," was closely linked to the first, since reformers also believed that community meant smallness. This movement made its debut in the middle schools as a way to reduce size and humanize or personalize learning (Rosario, 1981). Large middle schools were restructured around "houses," clusters, or communities to make up "schools within schools." While the guiding premise of this communitarian strategy was that school reorganization would make for more compassionate, caring, and community-like learning environments, the reform in many cases was adopted as a management tool for regulating student conduct. Community became an easy road to social control rather than an ethical blueprint for the manufacturing of a shared sense of a common good (Rosario, 2000).

At the high school level, the push for small schools came out of the work on "essential schools" pioneered by Theodore Sizer (1984, 1992) and others, for whom the principle "less is more" became the mantra as they struggled to rid America of the legacies of the "shopping mall high school" (Powell, Farrar, and Cohen, 1985)—such as curriculum differentiation, intellectual mediocrity, super-size instruction, and underachievement among minority youth—and transform high schools into more rigorous and manageable learning institutions (Sizer, 1984, 1992). CPESS was among the first schools in the nation to join the essential schools network in a progressive effort to produce "high schools on a human scale" (Toch, 2003), and to demonstrate that inner-city schools could work, that by reducing size, raising expectations, personalizing instruction, and trusting the power of ideas city children would achieve (Meier, 1995, 2002; Fine and Somerville, 1998).

At the time of filming, CPESS was unique among small schools in its organization and pedagogy. The school was a seventh-twelfth grade school divided into three divisions: Division I, serving seventh and eight graders; Division II, ninth and tenth graders; and Division III or Senior Institute, serving eleventh and twelfth graders. Divisions I and II were in turn divided into two houses, with each house having two teams: a humanities/social studies team and a math/science team. There were four team teachers in each house responsible for about eighty students, who stayed with their teams for two years. For those two years, students also belonged to a small Advisory composed of no more than 15 students. All school personnel were expected to serve as advisors.

In pedagogy, the school's codirector summarizes one of its key principles as follows:

> One of the ways that we've thought about the school is that if kids have developed five habits of mind . . . five ways of thinking about things . . . they should be able to graduate from here . . . think of perspective, from whose

perspective is something being presented? . . . What's your evidence . . . ?
. . . How's the thing that's being presented connected to other things? What if things were different? . . . And who cares? (Grant, 2006, p. 219)

In the film's last scene, the director expands on the rationale behind these habits as she explains to some visitors the school's organization. Comparing her experiences as a kindergarten teacher to the "powerful stories" she had heard about how doctoral work was managed at Oxford and Cambridge, she describes how the school is built on the educational principles undergirding these "two extreme ends of the educational spectrum." It is at these two ends, she says, "where we concern ourselves with . . . the individual's own interests, learning style . . . how to engage them. In which we respect their independence, their capacity to take on serious responsibilities." So CPESS was deliberately designed to preserve "the spirit of a good kindergarten" and an excellent university. In both could be found "what a good life is," people having "interesting . . . important . . . powerful conversations . . . the heart of what a powerful citizenry and a powerful democracy would be . . . citizens . . . in a position to carry out the public dialogue about the nature of their society, its purposes, and ways in which it could be improved." That was the key aim behind CPESS: "to design a school that will encourage that kind of habit of mind" (Grant, 2006, pp. 323–24).

High School II is a long film, and there are more than ample scenes illustrating how democratic logic is deployed to regulate the intellectual and moral orders of CPESS. Students at CPESS are not governed by force or authoritarian means; they are not expected to perform simply because an adult has ordered them to do so. The social contract at play here constructs them as morally autonomous selves. Teachers and administrators work to position students to view themselves as authors of their own lives, and they go about facilitating those constructions not through policing but through engagement in a collective process of community building where due regard and respect for oneself and others are guiding principles. I call this process of governance *soul making through community*. It is a notion whose derivation I owe to Rose (1999), who uses the term "government through community" to describe what happens when "in the institution of community, a sector is brought into existence whose vectors and forces can be mobilized, enrolled, deployed in novel programmes and techniques which encourage and harness active practices of self-management and identity construction, of personal ethics and collective allegiances" (Rose, 1999, p. 176). Rose's notion draws attention to how knowledge of the individual as encumbered, as composed of "bonds of obligation and responsibilities for conduct," becomes a resource in the fabrication of character. Individuals need not be governed by forcing compliance to external authorities; they need only be affirmed as individuals in their own right and invited to participate in fashioning their own identities as their communal ties and life histories are recognized and harnessed to buttress the process.

Two scenes in the film (3 and 17) that illustrate how parents are drawn into regulating their children's conduct offer examples of this type of governance. It is important to note at the start the absence in these scenes of the kind of power plays and relations characteristic of soul making in the earlier film. The scenes are more about engagement and conversation than about overt manipulation and positioning. In one such scene (Grant, 2006, pp. 205–07), for example, the viewer finds the school's codirector conferencing with a student, his mother, and the student's advisor. At stake is the young man's future at CPESS. He has "mixed feelings" about the school and he is being reminded of the social contract that binds the school community: CPESS is "a high school that works best for those people who really want to be here." The student is cast as having "anger that is racial," and he concurs with the assessment. At the risk of sounding "prejudiced," he concedes that he would probably be more respectful of a black teacher "cause he's black." He also admits to experiencing such feelings "a lot of times." But he also appears to reassure his audience. He believes his feelings are similar to those "that every kid gets," and he doesn't "feel that all the time . . . 'cause you have to respect all people no matter what race they are." But the codirector remains unconvinced. While he acknowledges the anger as a "real feeling" for a young man to have, he suspects that "especially at times like this," the student feels racial anger "more than other kids." He still sees reason for "real concern," and it appears his advisor perceives the same.

Like the codirector, the advisor acknowledges and respects the young man's anger as "a real feeling." He also believes, however, that the student must "take some responsibility and control for [his] actions." He is concerned how the anger may be affecting learning in his classroom. But "I can learn in your class," the student tries to reassure him, because he likes him more than Bridget, whom he does not respect. He respects the other teachers but "not some of the time." His racial feeling, he insists, is not constant. "It just flashes through my mind some time," he says.

His mother "totally" disagrees with her son's view on tying respect to race and reassures the staff of what she has "told [her] son 'over and over'": "that it doesn't matter what color skin a person is." Having the same color as one's teacher is no guarantee the teacher will treat one fairly, and she expects her son "to be responsible for [his] actions" and "to learn to relate to people just because they're people." In her view, everyone bears "a little prejudice." But what is critical to remember about prejudice is "how we use it." "That's the most important thing, I keep telling him," she says. "I'm not saying he shouldn't have feelings," she continues, "but it's how he uses his feelings. That's what's important. All of us have some feelings towards something. I'm sitting in the chair, I have them also and I'm not going to lie. But it's the ways I use it."

How does one govern a young black man who harbors racial anger in a school composed largely of students of color and a white faculty? Does one cast the young man as a racist and order that he leave or accept his material conditions? He is, after all, there by choice. Or does one invite that young man into a public conversation about that anger in the hope that he can be engaged and persuaded to see its potential threat to the school ethos? It is clear in this scene that CPESS has opted for the latter in accordance to its egalitarian principles. Also clear is the school's decision to bring his mother into the discourse. But equally important and clear as well is the way the school has chosen to shape that discourse. The codirector's main and leading appeal is to the social contract into which the family willingly entered. The family chose the school, and, while he has a right to his feelings, learning, a core value in the school community trumps such feelings. The young man must cast himself as a responsible member of that community and come to control his conduct. While he has a special responsibility to make something of himself, he must also demonstrate concern for others. Although "real," racial anger is no basis for denying others the respect they justly deserve as members of a community. The school believes that and so does the mother.

Yet, what the codirector, the advisor, and even the mother seem not to hear and address in the discourse as representative voices of the school community is the young man's contradictions: his sentiments do not all seem to ride on race alone. Bridget, who can be seen in other parts of the film, appears to be African American or a person of color. Yet, he does not respect her, but only, it seems, because he likes her less than some other teachers. In certain cases, liking someone appears to override race. "It might be," as Debbie Meier, the principal of the school at the time, suggested to me, "that when he doesn't like a teacher or feels dissed by them he just 'sees' them as white" (Meier, 2006, personal communication).

Although the young man's audience may have failed to acknowledge his contradictions, they seem nevertheless closer to "saving" his soul at CPESS. It seems doubtful they could have peered into his interior self in the absence of public and engaging discourse.

The second scene (Grant, 2006, pp. 234–36) poses a different kind of regulatory dilemma. Here, the viewer finds a teacher, in alliance with another parent, this time a father, engaging a student on the question of expectations. For the teacher, it appears, the issue centers on the student's lack of responsibility for what the community of teachers in the school, not just Bridget, demand: "for [him] to have a sense of how [he's] doing in [his] classes." Not acceptable is asking a teacher how he is doing; that is a matter for himself to know.

Although he confirms and agrees with the teacher's assessment, the father shares a different concern: his son's "responsibility to be honest" with his father, and, "even more importantly to be honest with himself." Troubling the father most is that his son has yet to decide for himself to move beyond the "satisfactory zone," the "*B, B-* zone," and develop the "fire in the belly," "the nervous energy," or "will to win" required to achieve greatness.

The son is unambiguous in his reply to the father's and teacher's constructions. He, too, desires greatness. He wants "to do better than passing" and "to really do very good," to go "the highest [he] can go." But he has "to do it [himself," he says. He "can't have people say, got to do this right now." He has "got to try and encourage [himself]."

Once again the school and family message is clear. To succeed at CPESS, individualism must triumph. Students must respond in conformity with how they have been constructed: as members of a community that cares about and believes in their power as individuals to accept ethical responsibility for their lives, make something of themselves, and go on to flourish.

Absent at the end of *High School II* is the kind of poignancy the last scene of *High School* leaves behind. To find such poignancy, the sort that accentuates the kind of student CPESS labors to produce, one needs to look to an earlier scene in the film. In this scene (32), a student unexpectedly rises to address the whole student body, which has assembled to listen to a choir composed of white high school girls from Michigan. Aware of the demonstrations Senior Institute students had planned for that afternoon in response to the Rodney King verdict, he shares the following:

> Y'all can hear me? Well, I chose upon myself to come here and stand in front of stage and talk to all y'all Central Park East Secondary School students to talk about . . . what we've been going through today. And I know from the Senior Institute that a lot of students are upset about what's been happening in Los Angeles. . . . I just want to tell them that nobody here's our enemy and . . . we have to stick together and . . . these people from Michigan . . . what they're doing here they're doing here for us. They're not here . . . to make us feel better, they're here because they like to sing and they want to show what they got. And they're not our enemies either . . . they're nobody in this room that are enemies and . . . doing anything drastic would just put us in a bad position and make us just like any other high school in New York City, which we're not. And, we just stick together and just stay with each other and not become enemies and show these people that we're not, we're not falling apart like other high schools in other states . . . I know . . . Senior Institute is real upset . . . but we have to do things peaceful . . . Showing your anger to these people . . . isn't going to do nothing for none of us . . . (Grant, 2006, pp. 293–94)

Here is no passive student, like the sort Wiseman claims to have found in Northeast High. His right to address the assembly has been exercised as the free individual the school construes him to be. But that seems to be as far as he would allow his individualism to go. His call for the renunciation of violence and for peaceful demonstration are not made on individualist grounds. He speaks not for himself but on behalf of community, his school, and its values. Here, it seems, communitarianism appears to be trumping individualist posturing—in keeping with the other competing principles the school advances.

The Shape of Things to Come: Governmentality in the New High School

The transition from the form of *soul making* displayed in *High School* to that displayed in *High School II* points to a larger transition in regulation that, citing Rose (1999), I described as the shift from discipline to "community" (p. 176). It is a change from an overt form of governance that occurs through the direct action at certain institutional sites such as the schools to a form of governance that occurs indirectly through the rules of reasoning and the discourses that penetrate and saturate those sites and are embedded in the cultural practices of everyday life. Control so understood is an all embracing feature of social life, embedded in and coterminous with its day-to-day ebb and flow (Deleuze, 1990). For Rose, it "is a moral field binding persons into durable relations. It is a space of emotional relationships through which individual identities are constructed through their bonds to micro-cultures of values and meanings" (p. 172). What is occurring in this transition is best captured in Foucault's notion of "governmentality" with its shift of attention away from state institutions and toward day-to-day cultural practices. According to Gordon (1991):

> State theory attempts to deduce the modern activities of government from essential properties and propensities of the state, in particular its supposed propensity to growth and to swallow up or colonize everything outside itself. Foucault holds that the state has no such inherent propensities; more generally the state has no essence. The nature of the institution of the state is, Foucault thinks, a function of changes in practices of government, rather than the converse. Political theory attends too much to institutions and too little to practices. (p. 4)

The institution that Wiseman depicts in *High School II*, then, offers a new form of regulation. Students are not directed or coerced from without by external authorities as they were at Philadelphia's Northeast High School. Rather, they are immersed in an environment that is saturated with the particular moral message of self-direction and commitment to common purposes and the common good that shapes and penetrates their very identities as they participate in the work of the school.

The soul making that occurs at Central Park East prepares students for a new and very different world than did the soul making that transpired at Northeast. This new world is a globalized one involving patterns of decentralized economic and political decision making that does not lend itself to the kind of direct and overt regulation of the state that has typified much of the twentieth century. Under these new conditions, the state has become an enabling institution that regulates indirectly through other agencies, particularly those within civil society. As Rose (1999) notes:

> The state is no longer to be required to answer all society's needs for order, security health and productivity. Individuals, firms, organizations, localities, schools, parents, hospitals, housing estates, must take on themselves—as "partners"—a portion of the responsibility for solving these issues—whether this be by permanent retraining for the worker, or neighbourhood watch for the community. This involves a double movement of autonomization and responsibilitzation. Organizations and other actors that were once enmeshed in the complex and bureaucratic lines of force of the state are set free to find their own destiny. Yet, at the same time they are to be made responsible for that destiny, and for the destiny of society as a whole, in new ways. Politics is to be returned to society itself but no longer in a social form: the form of individual morality, organizational responsibility and ethnical community. (pp. 174–75)

The locus of responsibility that once lay in such bureaucratic apparatuses of the state as the school is now shifted outward to numerous locations, public and private, in the larger society.

The state as a distinct and visible agency is to some degree displaced by the more diffuse entity of a set of common purposes and shared attachments that bind people together into a whole (Fowler, 1991, 1995). What is common and shared among proponents of this view is the notion of responsibility. The rights that are typically bestowed on individuals by the state become under this new arrangement part of one's individual responsibility. The various components of society's safety net, welfare payments, health services, unemployment insurance, and even education cease to be entitlements and instead require that individuals commit themselves to becoming self-sufficient. That is, they seek work, accept employment opportunities, and engage in the necessary learning that will maintain and strengthen their economic viability (Franklin, Bloch, and Popkewitz, 2003).

Such conditions require a new form of high school that is different from such large comprehensive institutions as Northeast that were modeled on the early twentieth-century factory. In its place, Central Park East provides a smaller and more intimate environment where students can engage in self-directive and collaborative work that cultivates the kind of stakeholder buy-in that is necessary for life in a globalized world. It is small schools such as Central Park East that hold the key to the shape of governmentality in the new American high school.

Yet, it is important to note that Central Park East is not a typical high school. In some ways, Northeast is closer, at least in size, organization, and pedagogy, to the high schools that populate the contemporary landscape. But there is an increasing effort among those who manage these large high schools to strike a balance between contemporary demands for accountability with the student-centered atmosphere of Central Park East (See, e.g., Kliebard and Stone, 2002). In a sense, we may be in the midst of changes that will ultimately transform the comprehensive high school into a more supportive environment that is more constitutive of academic success than what exists today.

References

Appiah, K.A. (2005). *The ethics of identity*. Princeton, NJ: Princeton University Press.
Benson, T. and Anderson, C. (2002). *Reality fictions: The films of Frederick Wiseman* (2nd ed.). Carbondale, IL: Southern Illinois University Press.
Cusick, P. (1983). *The ideal of egalitarianism and the American high school: Studies of three schools*. New York: Longman.
Dworkin, R. (2000). *Sovereign virtue: The theory and practice of equality*. Cambridge: Harvard University press.
Deleuze, G. (1990). *Negotiations*. New York. Columbia University Press.
Fine, M. and Sommerville, J. (1998). *Small schools, big imaginations: A creative look at urban public schools*. Chicago: Cross City Campaign for Urban School Reform.
Foucault, M. (1991). Governmentality. In G. Burchel, C. Gordon, and P. Miller (eds.), *The Foucault effect: Studies in governmentality*. Chicago: The University of Chicago Press.
Fowler, R.B. (1991). *The dance with community: The contemporary debate in American political thought*. Lawrence: University Press of Kansas.
———. (1995). *Religion and politics in America: Faith, culture, and strategic choices*. Boulder, CO: Westview Press.
Franklin, B.M., Bloch, M.N., and Popkewitz, T.S. Educational partnerships: An introductory framework. In B.M. Franklin, M.N. Bloch, and T.S. Popkewitz (eds.), *Educational partnerships and the state: The paradoxes of governing schools, children, and families* (pp. 1–23). New York: Palgrave Macmillan.
Gordon, C. (1991). Governmental rationality: An introduction. In G. Burchel, C. Gordon, and P. Miller (eds.), *The Foucault effect: Studies in governmentality* (pp. 1–51). Chicago: The University of Chicago Press.
Grant, B.K. (1992). *Voyages of discovery: The cinema of Frederick Wiseman*. Urbana, IL: University of Illinois Press.
———. (2006). *5 Films by Frederick Wiseman*. Berkeley, CA: University of California Press.
Grant, G. (1988). *The world we created at Hamilton High*. Cambridge, MA: Harvard University Press.
Handelman, J. (1970). An interview with Frederick Wiseman. *Film Library Quarterly* 3 (3): 5–9.
Jackson, P.W. The school as moral instructor: Deliberate efforts and unintended consequences. *The World & I* (1988) (March): 593–606.

Kliebard, H.M. and Stone, C.R. (2002). One kind of excellence: Ensuring academic achievement at LaSalle High School. In H.M. Kliebard, *Changing course: American curriculum reform in the 20th century* (pp. 107–25). New York: Teachers College Press.

Krupnick, C.G. (1995). *High school II film study guide.* Cambridge: Zipporah Films.

Levin, R.G. (1971). *Documentary explorations: 15 interviews with film-makers.* Garden City, NY: Anchor Press.

Lucia, C. (1994). Revisiting high school: An interview with Frederick Wiseman. *Cineaste* 20 (4): 5–11.

Meier, D. (1995). *The power of their ideas: Lessons for America from a small school.* Boston: Beacon Press.

———. (2002). *In schools we trust: Creating communities of learning in an era of testing and standardization.* Boston: Beacon press.

———. (2006). Personal communication.

Merelman, R.M. (1984). *Making something of ourselves: On culture and politics in the United States.* Berkeley, CA: University of California Press.

Merz, C. and Furman, G.C. (1997). *Community and schools: Promise and paradox.* New York: Teachers College Press.

Popkewitz, T.S. (1998). *Struggling for the soul: The politics of schooling and the construction of the teacher.* New York: Teachers College Press.

Powell, A.G., Farrar, E., and Cohen, D.K. (1985). *The shopping mall high school: Winners and losers in the educational market place.* Boston: Houghton Mifflin.

Rosario, J.R. (1981). *Mechanisms of continuity: A study stability and change in a public school.* Ypsilanti, MI: The High/Scope Educational Research Foundation.

———. (2000). Communitarianism and the moral order of schools. In B.M. Franklin (ed.), *Curriculum and consequence; Herbert M. Kliebard and the promise of schooling.* New York: Teachers College Press.

Rose, N. (1999). *Powers of freedom: Reframing political thought.* Cambridge: Cambridge University Press.

Rosenthal, A. (1971). *The new documentary in action: A casebook in film making.* Berkeley, CA: University of California Press.

Sizer, T.R. (1984, 1992). *Horace's compromise: The dilemma of the American high school.* Boston: Houghton Mifflin.

Toch, T. (2003). *High schools on a human scale: How small schools can transform American education.* Boston: Beacon Press.

Wiseman, F. (1968). *High school.* Cambridge, MA: Zipporah Films.

———. (1992). *High school II.* Cambridge, MA: Zipporah Films.

Chapter 7

The End of the Comprehensive High School? African American Support for Private School Vouchers

Thomas C. Pedroni

> *When they say, "What is happening to the kids who are being left behind [in public schools after vouchers]?" I say, "Why didn't you ask that question when you left them? Where was all the outrage when the people with money left us?" They say that the public school is where all the races and classes get together. In the Bronx? In the central cities of this country? I don't think so! The only people getting together there is us!*
>
> —Howard Fuller, Annual Symposium of the
> Black Alliance for Educational Options,
> March 2, 2001, author's fieldnotes

Do voucher programs signal the death of support among many African American families for the comprehensive high school or, for many working-class and poor African American parents, were reports of the comprehensive school's birth in the cities of the United States greatly exaggerated? Should we take the popularity of market-based educational reform among urban residents of color as the epitaph of the vision of *Brown*, or does it bespeak a reanimation of *Brown's* idealism within the post-welfare state? Have African American voucher parents lost interest in the promise of the comprehensive high school, or do they herald vouchers as the vehicle that might finally deliver some of the more valued elements of that promise?

This essay seeks to address these questions through analysis of ethnographic work with African American working-class families in Milwaukee, Wisconsin, who use vouchers as a means of removing their children from public secondary schools that they perceive to be unacceptable. In analyzing the ethnographic components, I propose to sketch out the points of overlap and departure among the educational visions of three sets of stakeholders within the debate around market-based reform of urban education: African American voucher families, the more powerful social forces that fund market-based reform in the United States, and defenders of the comprehensive school vision. This essay will have been successful if it helps defenders of the comprehensive high school envision strategies for incorporating the legitimate educational concerns of urban voucher families of color into more effective, meaningful, and democratic educational reform in the public sector.

The Historical Struggle for Quality Education for Communities of Color in Milwaukee

Those who defend the comprehensive high school in the United States have missed something essential in their inattention to the considerable support that market-based educational reforms, including vouchers, have received from marginalized urban communities of color. While equity-minded educational researchers have conducted studies demonstrating the particularly negative impact of educational marketization on the disenfranchised (e.g., Lauder and Hughes, 1999; Whitty, Power, and Halpin, 1998), not enough attention has been paid to the meaning of the crucial role the educationally dispossessed have actually played in building such reforms.

Locating the roots of minority support for market-based educational reform in previous historical struggles for educational equality, African American scholar and longtime voucher advocate Howard Fuller has heralded school vouchers as a key component of a "new Civil Rights agenda" (Fuller, 1985). For Fuller, voucher movements are a response to the failure of desegregation efforts to secure minority access to quality education and educational self-determination.

How should defenders of the comprehensive school read the pivotal role that urban community leaders and families of color have played in movements for vouchers in places such as Milwaukee, Cleveland, and Washington, DC? Are they the newest devotees of educational segmentation, privatization, and the free market fundamentalism of Milton Friedman (Friedman, 1955)? Have they given up on some of the best elements of the comprehensive high school partially enshrined in *Brown*?

The original vision of the comprehensive high school, I argue, contains elements that, given the narratives of African American voucher families in Milwaukee that I collected and analyzed, presents significant appeal at the same time that it inspires some detractions. As outlined in the *Cardinal Principles*, the comprehensive high school was to have both a *specializing* and a *unifying* function that together, it was deemed, would best reinforce the mutual benefit inherent in the relationship of the citizen to society (Wraga, 1994, pp. 23–24). New secondary schools available to all Americans, and not just the children of the elite, would allow both for the specialization of vocation that any society required, at the same time that they enabled the intermingling of all classes and races as preparation for citizenship in society. The comprehensive high school held out the promise that all young adults would be prepared both for civic life within an increasingly pluralist democracy and a productive work life within an increasingly complex economy (Krug, 1964, 1969; Labaree, 1988).

But, as I show, this vision of the comprehensive high school, despite its potential appeal, is not the point of debate in the current moment for African American voucher advocates. Although their narratives provide evidence that they might in fact find aspects of the comprehensive high school problematic, it is not the problem. Rather, any investigation of African American participation in voucher reform and its implications for the comprehensive high school must radiate from an honest appraisal of the conditions, contemporary and historical, that have characterized African Americans' experiences of schooling in the urban centers of the United States.

To what degree are the public schools against which African Americans have agitated both historically and in the current moment a reflection of their antipathy toward the goals enshrined in the comprehensive high school? Although in the following pages I sketch historical conditions and educational movements as they have existed in Milwaukee, the general historical outline, as well as the ongoing dissatisfaction of communities of color with the quality of urban public high schools, is likely to parallel experiences in other urban centers.

Milwaukee's communities of color have engaged in a long history of struggle for equal access to a variety of public goods and services, and movements to secure quality education for their children have been primary among these. The move toward support of vouchers has come only after a very long history of struggle for greater responsiveness from the Milwaukee Public Schools system (Rury and Cassell, 1993).

Beginning in the Civil Rights era, African Americans in Milwaukee participated in extensive direct and legal action to bring about the desegregation of their school district. Prior to a 1979 consent decree mandating desegregation, Milwaukee's history of segregation included a very elaborate and intentional system of unequal partitioning of resources, teachers, and students between predominantly white and predominantly black schools in the urban core. The essential priority of this system was to maximize educational quality for students of European American descent (Carl, 1995, p. 176; Fuller, 1985).

Predominantly black schools, even in times of exceptional overcrowding, were called upon to take responsibility for new black students, even when predominantly white schools in the area were noticeably undersubscribed. In the most extreme cases, a system of "intact busing" was devised, in which whole classrooms of black students from overcrowded "black" schools were transported by bus to undersubscribed "white" schools so that they might utilize separate classroom space there. The students of "intact busing" would report in the morning to their "home" school, board the bus for the predominantly white school, and return to their "home" school for lunch (at least until 1964) and again at the end of the school day (Carl, 1995, pp. 177–78).

Desegregation sought to end practices such as intact busing, and sought to bring about a redistribution of educational resources that would guarantee access to quality education for all students regardless of race. Yet the legacy of desegregation in Milwaukee is also a highly tainted one, as many studies have revealed (e.g., Dougherty, 2004; Fuller, 1985). In the hands of white politicians and school officials, the primary aim of Milwaukee's desegregation efforts eroded from guaranteeing educational opportunity to African American students into a superficial compliance to the desegregation decree, one that actually maximized the benefits of the desegregation system for white students. Funding formulas rewarded "white" schools both in the city and in the suburbs for taking on black students, who typically took long bus rides to school only to be separated from white students through systems of tracking. At the same time, public historically black neighborhood schools in the inner city were closed down in order to make way for specialty magnet schools (Metz, 2003) which, although public in name, engaged in admissions practices that made them into overwhelmingly white institutions. Tellingly, while desegregation meant the busing of some black children into predominantly white schools in predominantly white neighborhoods, and the busing of white students to newly formed magnet schools from which neighborhood black children were largely excluded, it never involved the busing of white students into predominantly black schools in the inner city (Dougherty, 2004; Fuller, 1985).

Rather, Milwaukee's "forced busing" program for students of color, as many African Americans have called it, resulted in a tremendously costly and baroque transportation system. In perhaps the most extreme example, black children from what was previously a single neighborhood school's catchment area were bused to 97 separate schools throughout the city of Milwaukee (Fuller, 1985).

Not only has this program been criticized for its tremendous inefficiency in utilizing educational resources for educational benefits, but it has also been decried for the enormously destructive effects it has had on black students, their families, and the black community in general (Fuller, 1985). First, black students, unlike most of their white counterparts, endured long bus rides twice a day to and from their home neighborhoods, sometimes as long as an hour or more each way. Busing also made involvement in their children's schools an insurmountable challenge for many black parents and families, as visiting the school their children attended now required a substantial journey across town into an unfamiliar and oftentimes unwelcoming neighborhood. This proved to be particularly difficult for African American households in which no one owned a car, and in which all adults worked away from home.

The failure of desegregation as it was actually carried out in Milwaukee to adequately address issues of educational quality for black students, coupled with the closing of many predominantly African American neighborhood schools, has historically resulted in the creation of a movement for schools controlled by Milwaukee's communities of color. The decade after the Civil Rights era saw the birth of a number of black-controlled independent private schools (many of which still exist today) that have historically sought public funding (Carl, 1995, pp. 248–49). Beginning in the mid-1980s many African American community leaders participated in a narrowly defeated effort to create a predominantly black public school system out of the set of remaining public schools around North Division High School on Milwaukee's north side (Carl, 1995, pp. 240–43).

Coupled with the reality of a political climate of insurgent conservatism in Wisconsin, as well as a relative increase in black political representation in Milwaukee, the continued frustration of communities of color with what they saw as the public school system's intransigence paved the way for Milwaukee in the late 1980s to become, with the critical assistance of conservative grant-makers such as the Bradley Foundation, the staging ground for the first modern-day voucher experiment in a large urban area in the United States (Carl, 1995, pp. 255–95).

It is within this post-*Brown* context of persistent and grave educational inequality that African American investment in voucher programs, and the implications of this for their support of the comprehensive high school, should be read.

Listening to African American Voucher Families

Public school has a lot of changes that I felt that needed to be made. I'm not knocking public schools. Public school has a lot of good things to offer. But public school also on the other hand has a lot of improving to do. And I resented that being African-American—and of course I live in one of the poorer neighborhoods—my children were stigmatized by that. And they felt like they were giving you something. I'm a working mother. I pay taxes. It's like nobody else's . . . . And my taxes help pay for public education. So as far as I was concerned, it was a paid education. You know, and I didn't appreciate the stigma like you have to take whatever I give you, you know. It's free. You ain't paying for nothing. And you know, that was the stigma. And it was so hard to get anything done. I was always . . . . It was always a fight. And I was looking in search of something different.

—Sonia Israel, Mariama Abdullah School
voucher parent, quoted in Pedroni, 2004, p. 156

The argument to which I now turn in this essay is that working-class and poor families of color use vouchers and become allied with an otherwise conservative educational movement as a result of the "elements of good sense" (Apple, 1996; Gramsci, 1971) they possess concerning the poor quality of education available to African American children through urban public schools. This movement to vouchers, like other historical struggles for quality education on the part of Milwaukee's communities of color, is a product of agency on a terrain not of parents' own choosing. In the current debate over privatization as a prescription for urban public school woes, I want to argue that African American investment in vouchers is tactical, and is not the result of a primary commitment to educational free markets as a solution to social ills, nor to a subsidiary vision of educational markets as a mechanism for delivering a differentiated array of "educational products" to consumers with differing educational needs and tastes. African American alignment with conservative educational forces in American society, I argue, is thus highly conditional and fleeting, and does not in and of itself constitute a rejection of the comprehensive high school.

In undertaking such an analysis I conducted ethnographic work with families who utilized vouchers to enroll their children in two private voucher high schools in Milwaukee. Half of the families enrolled children at a Catholic High School for young women, and half in a nonsectarian vocational school for "at risk" youth. These two schools were selected as part of a larger study I have conducted around race, subaltern agency, and identity formation among voucher families in Milwaukee (Pedroni, 2004, 2005; see also Apple and Pedroni, 2005).

I coded transcripts of interviews with families at the two schools for themes related to the primary questions of the larger study. First, I coded for autobiographical information regarding participants' life histories as well as their children's scholastic histories. Next, I coded participants' remarks on the positive and negative attributes of their chosen voucher schools and the voucher program (the Milwaukee Parental Choice Program) in general. Finally, I catalogued comments from parents and guardians that reveal their perspectives on rejected public schools attended by their children and Milwaukee Public Schools (MPS) in general. While my original intention in conducting these interviews was not to assess respondents' dispositions on the comprehensive high school (this was not a question that was central to the larger study), I hope to demonstrate that their narratives regarding their voucher advocacy did not for the most part entail a preference for resegregated schools and greatly differentiated school missions, nor a rejection of most facets of the comprehensive high school vision, as is often argued. Although the rhetoric of neoliberal supporters of educational marketization would have us believe that vouchers promote diversification of school missions and practices, and the rhetoric of critics of educational market reform would have us believe that vouchers necessarily imply resegregation, I intend to demonstrate that vouchers do not promote, nor do African American voucher advocates primarily want, either of these things.

In what follows, I present the remarks of parents enrolled at the two schools regarding MPS and Milwaukee Parental Choice Program (MPCP) schools "at face value." That is, my purpose is not to discern whether the schools and the programs actually function and perform the way the parents I interviewed said they do. (In fact, during field visits with personnel at the schools it became clear to me that, while some characteristics of the schools seemed to match my interview subjects' portrayals fairly well, others most certainly did not.) Instead, I am primarily interested in the consciousness and ideological processes that are in motion as families navigate their choices within Milwaukee's "educational market." Because I present their remarks at face value, and because I received help from voucher school administrators in identifying families to interview, we can expect their comments regarding their chosen schools to be very positive. Conversely, their comments about schools they've rejected can be expected to be quite negative.

While this approach doesn't necessarily give us reliable information about the various public and private schools interview participants discussed, it does give us a fairly good sense of their ideological formation in relation to vouchers and voucher schools. This in turn helps clarify why voucher families choose particular schools while rejecting others. Even more importantly, it reveals how, in what ways, and to what degree they become allied with more free-market oriented voucher proponents or reject the central organizing principles of the comprehensive high school.

In the presentation and analysis of fieldwork that follows, all names of individuals, schools, neighborhoods, and other explicitly identifying factors in the data have been altered. This is undertaken to protect individual and institutional identities from potential harm.

Saint Urbina Catholic High School

Saint Urbina, which identifies itself as a "multicultural immersion school" for young women, has played a role in educating children living in

Milwaukee's urban core for half a century. The school currently serves 325 young women in grades 9 through 12, of whom 60 percent receive vouchers through the Milwaukee Parental Choice Program. According to its mission statement, the school is committed "to maintaining diversity in ethnic, religious, cultural and economic backgrounds of students." Today, the racial and ethnic makeup of the school, according to the school's website, is 33 percent European-American, 27 percent Latina, 26 percent African-American, 6 percent Southeast Asian-American, 5 percent Middle Eastern-American, 2 percent Native American, and 5 percent multiethnic, multicultural, and/or multinational.

Religiously, the school's composition is reported as 62 percent Catholic and 38 percent non-Catholic. While the school bases its curriculum in Catholic teachings, it also maintains, according to its website, a strong commitment to Gardner's theory of Multiple Intelligences in curriculum, pedagogy, and evaluation (Gardner, 1993). Furthermore, St. Urbina emphasizes college preparatory and business preparatory educational tracks. Ninety-one percent of the school's young women enter college. On more than one occasion the principal of St. Urbina pointed out to me the ways in which, therefore, St. Urbina more closely approximates the vision of an integrated high school (although it is segregated by sex), also accommodating more career trajectories than the public high schools in the area.

Dasha Dapedako

Dasha Dapedako is an award-winning jazz vocalist, a renowned African storyteller, and a paraprofessional fifth-grade reading teacher at a Milwaukee Public Schools elementary school. Her daughter, Namiya, is a high school senior at Saint Urbina, in which she enrolled as a ninth grader after attending public schools at both the elementary and middle school level. Although Namiya had not wanted to attend the all-girls Catholic School, preferring instead the more social environment of the public school that her friends would attend, Ms. Dapedako insisted on St. Urbina. Enabled by the recent availability of vouchers for sectarian school attendance, Dapedako removed her daughter from a school where she felt her daughter was likely to affiliate with people who were not positive influences. "I think [St. Urbina's] really saved her, because she's the kind of kid that's easily led."

Primarily, what creates such concerns for Ms. Dapedako is school size. "It's too many kids. It's just too many kids in one building. . . . And the class sizes, unless they're on SAGE [a class-size reduction program] in the lower grades, the class sizes average 26, 27, 28, 29. That's a lot to handle . . . And anything over 20 is too much." To the degree that a public comprehensive high school entails a large school with large class sizes, Ms. Dapedako could be considered a critic of that vision.

But for Dapedako, school and class size is not the only element that attracts her to St. Urbina. Besides offering substantially smaller class sizes, Dapedako was attracted to Saint Urbina's by a host of other qualities. According to Dapedako, most public schools in Milwaukee suffer from cultural and socioeconomic homogeneity. "You have mostly kids that come from the same cultural base, same economic base, and you know the same modern slang teenagers. And that's the paradigm in front of them and that's what they perpetuate." This is something which she feels contrasts markedly with the composition of St. Urbina's student body. Whereas at most public schools, students "don't feel any grounding, don't feel any closeness," at her new school, largely because of this diversity, Namiya has "adopted very quickly a sense of family." At St. Urbina's, the diversity that exists is seen by students and faculty to be a very positive attribute of the school.

> She adopted all the girls as her play sisters—the Arab girls, the Spanish girls. She brought home so many kids of so many different ethnic backgrounds, which was wonderful. And they experienced us, and we experienced them . . . . And it was a true exchange. It wasn't just a brush with. I mean, they actually talked about the problems that they had adjusting to this society, and she talked about what it was like being African-American. I mean, they actually exchanged and I think that's more value. You couldn't get that out of the book. That kind of experience is had, it's not read.

While the intimacy within the school setting that Dapedako valorizes in St. Urbina's does not necessarily reflect the more impersonal ethos of the comprehensive high school, clearly the rejection of homogeneity and the preference for the value of a more demographically heterogeneous population does.

Gina Price

Gina Price is a single parent pursuing the first two years of a degree in education through Milwaukee Area Technical College. In addition to her studies, Ms. Price is a teacher in a first grade elementary school. She also works two other part-time jobs—one with the school's after-school program, and the other in security. "It takes a lot of energy."
Her daughter, Patricia, decided to apply to St. Urbina High School already in middle school. Price agreed with her daughter's decision, but would have also been satisfied with a few public schools that she had been considering.
However, due to overcrowding, Ms. Price feels that public school teachers are often not able to tailor their teaching to individual students' needs. "In the public school . . . you don't get around to all the children . . . . [Teachers]

can't just say, 'Okay. I know he needs this so I'm going to work with him on that.'" Here it is clear that Price's concern is not with the public school's mission per se, but rather with the overcrowded conditions that prevent teachers from having the effect on students that they would have if they were given the kind of conditions she and her daughter find at St. Urbina.

Price worried about the distractions posed by what she describes as a daily fashion show at public schools. "You see the people with the name brand clothes, and children taking other children's things." At St. Urbina's, because of uniforms, her daughter does not have to contend with such distractions. "Everybody wears the same thing. You don't have to worry about, 'Oh, your mom bought you that, and this is Kmart and this is Wal-Mart.'" Price also appreciates the strict discipline and academic seriousness that she feels many private schools are able to enforce because of their ability to use expulsion as a threat. "If you don't pick up that grade point, they'll take you out . . . . They'll tell you, they won't put up with that foolishness . . . . There's strict rules and strict policies . . . . They've had seven people's children get kicked out because of . . . retaliating." She is intimately familiar with the ability of MPCP schools to carry out such expulsions, because many of those removed from voucher schools end up in her classroom. "We've got more children coming out of voucher schools, because of behavior . . . . We've gotten like at least four to five in our class alone. And then after a while you don't have to wonder about why they're out."

Jan Lincoln

Jan Lincoln's daughter transferred to St. Urbina earlier this year as a tenth grader from a public high school. The decision to enter the all-girls Catholic High School was almost entirely her daughter's own choice.

> It was her choice. She decided that she wasn't learning at [the public high school]. She said she thought it was like kids were too rowdy. And . . . the teachers couldn't help everybody, or her thing is they *wouldn't* help. And she was like, the classes were too big, and it was like a fashion show. She couldn't learn, you know, being like, them being into boys, and you know, she just decided.

Lincoln's daughter is determined that she will go to college, and would like to become a lawyer. She initially learned about St. Urbina from her two older cousins, who already attended the school. Once she was enrolled, "her grades came up. She's learning more, she comes home from school, do homework, do projects. I mean it's like, she a whole new person since she had been to [St. Urbina]. I mean, grades came up real extremely high . . . . And she had no problem getting up in the morning."

According to Lincoln, this behavior was in stark contrast to her daughter's year at the public high school. "I left to go to work, like at 5:30 in the morning, and then I called to get them up. She just felt, she had a headache, she ain't going to school. . . . So she missed a lot of days of school."

For Lincoln's daughter, St. Urbina's school offers a career trajectory that was only rhetorically available at the public school. Whereas St. Urbina succeeds in making college a practical choice for its students, the limitations of the public school did not in practice enable this trajectory. By default, the crisis conditions of most urban public schools impose a narrow curriculum centered on discipline and control, rather than a differentiated curriculum enabling multiple career trajectories. Ironically perhaps, the everyday reality of St. Urbina more closely approximates the comprehensive high school vision than the more segregated population and more homogeneous curriculum offerings of the public high school that Lincoln's daughter rejected.

Lincoln is satisfied that St. Urbina has had much higher expectations of her daughter. "When you miss a day of school, your grades drop. And she's going to school every day. She don't want to get no low grade." Furthermore, "They challenge her, because they're much stricter. They want you to learn. They want you to get good grades. And she see that. She like it."

Knowledge Ventures Learning Academy

Knowledge Ventures Learning Academy is a vocational and academic high school on Milwaukee's north side. According to its mission statement, Knowledge Ventures specializes in "addressing the educational needs of At Risk Youth, teen parents and Learnfare students."[1] The school gives preference to school age parents, and has a total enrollment of 208 students, of whom 175 (84 percent) receive vouchers. The majority of the non-MPCP students are behavioral reassignment students, enrolled at Knowledge Ventures through a contractual agreement with Milwaukee Public Schools. Ninety-seven percent of the school's students are African-American, and 58 percent are female.

Although the school offers a college preparatory track, it also provides vocational courses in cosmetology, home mechanics, day care certification, photography, food service, computing, massage therapy, sewing, and business management.

Samantha Murphy

Samantha Murphy, the mother of two children at Knowledge Ventures, is currently pursuing a degree in human services at Milwaukee Area Technical College. Her two children—an older son who graduated the previous June, and a younger son who is a senior this year—have each spent three years in attendance at Knowledge Ventures.

Murphy pulled her two teenage sons from public schools when she realized that "them being in a bigger environment wasn't working." Initially her children had attended public schools on Milwaukee's south side that Murphy found fairly satisfactory. But when the family relocated to a near north neighborhood, "they put [them] in a north side school closer to my house, and it was *not* going on." She explains, "Well everybody knows that the funding for north side schools are not the same as the south side schools."

The public school into which her sons were placed did not meet her standards in a number of ways. Everyday classroom life was rowdy, and she became dismayed by her children's attempts to "fit in" by assimilating into what she characterized as the school's academically and socially dysfunctional culture. Teachers would spend most of the time in the classroom attempting to discipline children. "By the time they get the classroom under control, it's time to go home, and move to the next class." The more intimate environment and individual attention of Knowledge Ventures Learning Academy, something not explicitly lauded by advocates of the comprehensive school, is certainly something that Murphy values.

Murphy also found herself deeply incensed by the level of suspicion with which the public school regarded her children. When it was discovered that her son had gotten a ride with another student at the school—in a vehicle that the other child had taken on a test drive without returning—the school's principal blamed Murphy's son equally for the alleged crime. Although her son had assumed the vehicle belonged to the young man's mother, the principal "told my son, in front of me, 'What would make you do something like that? Next time you do something, you're out of here. You're going to alternative.'" Murphy realized that, in a school environment in which the principal was quick to treat students as criminals, such attitudes were likely to flow through the entire faculty. "If the head is speaking into that, then everybody . . . you know, then . . . he's passing it down, you know what I mean?" Here Murphy is rejecting not integration, but rather the overtly racial diminution of her son.

Gary Johnson

As he openly explains, Gary Johnson is a former gang leader and ex-convict, and the father of two young women attending Knowledge Ventures Learning Academy. His primary objective with his daughters is to prevent them from making some of the wrong choices that he feels he made earlier in his life. And, according to Johnson, this is proving to be quite a challenge. "My youngest daughter has a behavior problem, and really the only person who can control her and contain her is me." He identifies a different sort of problem with his older daughter. "Her problem . . . was, she kind of like fell in love with this boy, so once he graduated, she really didn't want to go to school there."

Upon the advice of a cousin, Johnson moved his older daughter to Knowledge Ventures, where he had heard that children were eligible to work. "And so I'm like, okay, that's a foundation there." Although the promise of work turned out to be elusive, Johnson was amazed by the progress his daughter began making academically. "Right away her grades and everything just changed." He subsequently enrolled his younger daughter as well.

Johnson faults the Milwaukee Public Schools for their unwieldy size. "The classrooms are so big, they have so many students in them, there's no way that that teacher is going to get to know every student individually." At Knowledge Ventures, now that his older daughter is pregnant, he feels she receives the individualized attention and support that she needs within an educational context that Johnson refers to as "a good family." Again, this valorization of the interpersonal and the familial may not coincide with the core tenets of the comprehensive high school.

Johnson also applauds the ability of the teachers at Knowledge Ventures to make curriculum appealing, interesting, and relevant to students who have not always experienced schools as engaging places to be. "I believe that, like they say, the old saying goes, if you don't want a Black male to know anything, put it in a book, because they won't pick it up." At Knowledge Ventures,

> "They get creative, the teacher, to make them want to learn things." Particularly in relation to his daughter, "If she don't have any interest, I don't care how smart she is, she's not going to give you any of her knowledge, any of her input. And she's really interested in the school. . . . They make my kids want to come to school. When I take my [church-related] trips out of town, they don't want to go. They want to go to school."

Ramona Aguila

Ramona Aguilar works at her church's food pantry, and is the foster mother of four children currently attending four different Milwaukee Parental Choice Program schools. Ms. Aguilar's younger daughter transferred to Knowledge Ventures last year as a sophomore, in large part because of large class sizes. "The ratios are bigger, the class ratio to the teacher in public schools are much larger than at a small private or alternative." Under such overcrowded conditions, Aguilar wonders,

> How much time does it take a teacher to do roll call, limiting the instruction time? And then when you did have specific issues that you needed to talk to,

the teacher does not have time in a public school. At [her previous public school] . . . she had 30 other students amongst her . . . . So that was one of the reasons why [we] brought her here.

The overcrowded condition of her public school classroom frustrated the daughter to the extent that the values Aguilar had taught her at home—to respect her elders, for example—were being challenged and threatened.

Aguilar contrasts the attentiveness of teachers at Knowledge Ventures with the anonymity her daughter experienced at her public high school. She offers her experience with parent-teacher conferences as a case in point. At the public school's conferences, "They're all in the cafeteria, everything's set up in tables. You have 2 to 3 teachers per table. . . . Teachers have to look for the 300 students that they see all day, and figure out, what they can just briefly tell [me] about [my] daughter." In contrast, at Knowledge Ventures, "You have that flexibility to go into the teacher's classroom, meet with that teacher, and everything that is said to the teacher, there's no one else to hear it." Again, a smaller more intimate school environment is more valued than the impersonality of a larger setting.

Aguilar blames what she regards as the poor conditions in public schools on two factors: inadequate funding—"things getting cut all the time"—and the fact that "teachers get such a high salary sometimes, that they . . . it's the same if they teach, the same if they don't." She finds that this is particularly true of veteran teachers.

> Teachers that had been there 15 years or more were the ones that just didn't care, you know? It was worksheets, and . . . . no challenging. And so that classroom, you would go in there, you would have a lot of constant disciplinary problems. But yeah! You had disciplinary problems because the kids are bored and the teacher's fed up!

Aguilar attributes this teacher apathy not just to overcrowding, underfunding, and the disconnection between teachers' salaries and the success of their students. She also feels that such teachers are motivated by elitism. "Their class is higher than, you know, compared to the students' class." Lack of tolerance for internal differences—in socioeconomic status between teachers and students, as well as culturally based readings of orderliness (with these apparently more valorized among Latino families than white teachers) speaks to a perception of the voucher school environment as more accepting of the mingling of differences than the local public school.

While Aguilar is critical of many public school teachers, she also expresses sympathy for the way she feels they are overburdened. "They load you with so many things that, towards the end, you're tired. You don't feel like doing it no more because it's the overload. [You're] burnt out."

Implications for the Comprehensive High School

As these examples help illustrate, and as the larger study of which these interviews were a part revealed, families who utilize vouchers are relatively unified in their assessment of the qualities that lead them to reject certain schools while choosing others. For many working-class and poor families of color in Milwaukee, vouchers enable their children to escape schools beset by racially pathologizing discourses and overcrowded classrooms. Utilizing vouchers, parents and guardians are able to choose small schools with smaller class sizes, which they believe offer the individualized attention, differentiated curriculum, and situated disciplinary practices that best serve their children. In part, school practices such as these succeed, parents explained, because the flexibility that private schools can exercise enables them to root such educational practices within the culture, everyday lives, and social conditions of the families they serve.

Each of the parents and guardians at Knowledge Ventures, like the parents at St. Urbina, stressed above all the importance of small class size. In the end, according to the parents interviewed, what prevents public schools from having small class sizes is neither their unresponsiveness to parents, nor their market inefficiency, as market advocates for vouchers would have us believe. Clearly public school teachers would want smaller class sizes, according to the interviewed parents, and that which prevents public schools from having small class sizes, rather than market inefficiency or lack of market discipline, is the lack of the funding that would enable this. So do the qualities that voucher parents seek out in schools they choose fit more closely with market advocates or defenders of the public comprehensive school?

Not only is support among African American working-class and poor families not an endorsement of free-market principles of reform, but it also does not necessarily signal a rejection of the principles underlying the comprehensive high school. First, it must be acknowledged that for most urban public school students of color and their families, the dream of the comprehensive high school was never realized in any meaningful sense, due to factors including underfunding, segregation, and the ubiquitous presence of pathologizing racial discourses. If there has been a racially based rejection of the comprehensive high school, then it has been whites, and not blacks, who have overwhelmingly spearheaded this rejection, a fact to which the quote from Howard Fuller that opened this essay eloquently alludes.

Furthermore, while free-market enthusiasts have argued that the introduction of market mechanisms into education will produce greater variety in educational forms, and thus a tremendous increase in "market segmentation" in education—a direct contradiction of the more universalist essence of the comprehensive high school—much research, including my own, points to the conclusion that markets actually produce more, not less, educational standardization and conformity in school mission and curriculum (e.g., Benveniste, Carnoy, and Rothstein, 2002). Therefore, not only do black voucher advocates not endorse free markets as a first principle, but free markets are not essentially counterproductive to furthering the more universalist ethos of the comprehensive high school. In fact, as the discussion of St. Urbina in this essay exhibits, there is irony in the fact that many private schools (including, in Milwaukee, private voucher schools) actually better approximate the comprehensive high school than urban public high schools, the latter well known, at least to their majority clientele of working-class and poor families of color, for their tremendous homogeneity in both demographic makeup and in their meager academic offerings.

Given this, the task for supporters of the public comprehensive high school is to recognize that African American voucher families are not rejecting the comprehensive high school, but rather are rejecting a public high school that never attained anything like the stature that promoters of the comprehensive high school envisioned. Public urban high schools, certainly those intended for working-class and poor people of color, have never deviated much from being overwhelmingly segregated institutions offering a singular curriculum focused on the disciplinary control of the populations contained in their buildings. Although there is an implicit critique of the comprehensive high school built into the narratives of the parents I interviewed—that the comprehensive high school does not offer the more intimate size and sense of community and individualized attention that many vouchers parents crave—parents actually yearned for greater heterogeneity among both students and curriculum tracks than the public schools had to offer. That is they were rejecting some of the same qualities in the public schools that comprehensive school advocates would reject.

The same qualities that have made urban public schools unattractive to voucher parents have furthermore prevented those schools from attaining anything close to the stature that promoters of the comprehensive high school have envisioned. Whites, not blacks, through their flight to the suburbs, have rejected the value of racial integration in schools. The underfunding that has resulted from this flight has contributed to unwieldy class sizes as well as the narrowing of the curriculum. And the under training of low-paid teachers in skill sets and dispositions more appropriate for engaging students in urban public schools has resulted in a climate in which the remaining majority population itself feels that its cultural attributes are denigrated.

The lack of the conditions required for creating the comprehensive high school in urban public schools has spurred the interest in vouchers, and so it is not African American voucher families who are hastening the comprehensive school's demise.

Note

1. Learnfare is a component of Wisconsin Works (W-2), the state-created welfare entity formed in Wisconsin at the time that federal legislation under the Welfare Reform Act devolved most aspects of social welfare provision to the 50 states. Under Learnfare, parents who are placed in jobs funded through Wisconsin Works are fined if their school-age children are not enrolled in schools. Learnfare students meet regularly with "Case Managers," who, beyond monitoring enrollment and attendance, provide the following services: "assessment, career development and planning, problem solving and role play, non-traditional counseling, crisis counseling and intervention, supportive services, and referral to community services" (Wisconsin Department of Workforce Development).

References

Apple, M.W. (1996). *Cultural politics and education*. New York: Teachers College Press.
Apple, M.W. and Pedroni, T.C. (2005). Conservative alliance building and African American support of voucher reforms: The end of *Brown's* promise or a new beginning? *Teachers College Record* 107 (9): 2068–105.
Benveniste, L., Carnoy, M., and Rothstein, R. (2002). *All else equal: Are public and private schools different?* New York: RoutledgeFalmer.
Carl, J. (1995). *The politics of education in a new key: The 1988 Chicago School Reform Act and the 1990 Milwaukee Parental Choice Program*. Unpublished doctoral dissertation, University of Wisconsin, Madison.
Dougherty, J. (2004). *More than one struggle: The evolution of Black school reform in Milwaukee*. Chapel Hill, NC: University of North Carolina Press.
Friedman, M. (1955). The role of government in education. In R.A. Solo (ed.), *Economics and the public interest*. New Brunswick: Rutgers University Press.
Fuller, H. (1985). *The impact of the Milwaukee Public Schools system's desegregation plan on Black students and the Black community (1976–1982)*. Unpublished doctoral dissertation, Marquette University, Milwaukee, Wisconsin.
Gardner, H. (1993). *Frames of mind: The theory of multiple intelligences, 10th edition*. New York: Basic Books.
Gramsci, A. (1971). *Selections from the prison notebooks of Antonio Gramsci*. Trans. G. Smith and Q. Hoare. New York: International Publishers.
Krug, E.A. (1964). *The shaping of the American high school*. New York: Harper and Row.
———. (1969). *The shaping of the American high school, Volume 2*. Madison, WI: University of Wisconsin Press.
Labaree, D.F. (1988). *The making of an American high school: The credentials market and the Central High School of Philadelphia, 1938–1939*. New Haven, CT: Yale University Press.
Lauder, H. and Hughes, D. (1999). *Trading in futures: Why markets in education don't work*. Buckingham: Open University Press.

Metz, M. (2003). *Different by design*. New York: Teachers College Press.//
Pedroni, T.C. (2004). Strange bedfellows in the Milwaukee "parental choice" debate: Participation among the dispossessed in conservative educational reform. *Dissertation Abstracts International* 64 (11): 3946A. (UMI No. 3113677).

———. (2005). Market movements and the dispossessed: Race, identity, and subaltern agency among Black women voucher advocates. *Urban Review* 37 (2): 83–105.

Rury, J.L. and Cassell, F.A. (Eds.) (1993). *Seeds of crisis: Public schooling in Milwaukee since 1920*. Madison, WI: University of Wisconsin Press.

Whitty, G., Power, S., and Halpin, D. (1998). *Devolution and choice in education: The school, the state and the market*. Buckingham: Open University Press.

Wisconsin Department of Workforce Development. (2002). *Wisconsin Works (W-2): Learnfare*. Retrieved May 6, 2004 from http://www.dwd.state.wi.us/dws/w2/learnfare.htm

Wraga, W.G. (1994). *Democracy's high school: The comprehensive high school and educational reform in the United States*. Lanham, MD: University Press of America.

Comparative Perspectives

Chapter 8

The Formation of Comprehensive Education: Scandinavian Variations

Susanne Wiborg

The Uneven Development of Comprehensive Education

When looking at Scandinavian education it is not so much the question of whether comprehensive schooling is struggling to survive that comes to mind, as why comprehensive schooling is doing so well in this part of the world. The comprehensive school systems in Denmark, Norway, and Sweden are not, of course, without their own problems, but they are unique in having an almost universal public school from grade 1 to 9 or 10 with mixed ability classes. The primary and lower secondary part of the public school is integrated into one system of all-through education where selection to further education is postponed until the age of 15 or 16. In this chapter I wish, therefore, to bring particular attention to its unique historical formation in order to offer an explanation of this particular outcome.

The origins of comprehensive education in Scandinavia dates back to the late nineteenth century and was developed through three key historical events. First, the parallel school system, which consisted of an elementary school attended by children of the rural and urban working classes, and a secondary school (Grammar school), which enrolled children of the bourgeoisie, was transformed into a linear structure by the creation of a middle school. This provided a bridge between the elementary school and the secondary school, and the latter was thus converted from nine to three years and renamed the *gymnasium* in Norway and Denmark and *läroverk* in Sweden. The system now consisted of the elementary school, the middle school, and the upper secondary school in that sequence. At the same time, the private schools were reduced numerically, which meant that the elementary school could serve as a proper basis of the further education. This was a unique step in the history of Scandinavian education, firstly because most children, irrespective of social class, became increasingly enrolled in the public system of education and, secondly, since it allowed the children progressing through the system according to academic ability and aptitude. The middle school was introduced earliest in Norway, already in 1869, and later in Denmark, in 1903, and in Sweden in 1905.

The second event was the abolition of the middle school, which paved the way for a seven-year comprehensive school (with streaming in the top two classes) and the maintained three-year upper secondary school. This event happened in Norway in 1920, in Sweden in 1927, and in Denmark as late as 1958. After World War II the effort was concentrated on extending the seven-year comprehensive education to nine years in order for children to stay together in the same class for a longer period. This occurred in Norway in 1969, a little earlier in Sweden, in 1962, and again later in Denmark, in 1975. The third event, which occurred during the 1980s and the 1990s, was the introduction of mixed ability classes whereby streaming according to ability was finally abolished. However, pupils can today be grouped according to ability within the framework of mixed ability classes, but only for a limited amount of time during the school term. This is an expression of a general will to integrate mixed ability classes and ability groupings in a balanced way to avoid the enhancement of academic standards at the expense of social cohesion (Markussen, 2003; Skovgaard-Petersen, 1976; Dokka, 1967; Telhaug, 1994; Richardson, 1999; Wiborg, 2004).

This outcome—an all-through, nonselective system of education from grade 1 to 9/10 with mixed ability classes common to the Scandinavian countries—raises the question as to why it has been possible to develop a comprehensive school system to its full potential in these three countries when there have been such obstacles to this in most other European countries? (see Leschinsky and Mayer, 1999). This identification of a common outcome offers a possibility of comparative macro-historical analysis. I am seeking a single set of factors that the Scandinavian countries share in common—and that are absent in other countries—in order to develop a general explanation of why these countries went along the same track when introducing comprehensive education. Scandinavian school history has fostered numerous single-case studies that in detail outline the specific national development of comprehensive education. However, even though these studies have provided us with important contextualized accounts, they are nevertheless so deeply entrenched in national history that generalization beyond the particular is almost impossible. In order to explain why the Scandinavian countries have introduced radical comprehensive education simply requires a search beyond the particular for a single set of factors that determines this common outcome. It is important to stress, however, that this does not mean ignoring vital contrasts between the countries; on the contrary, these are in particular helpful in explaining the different timing of the introduction of comprehensive education. As we shall see later, comprehensive education was introduced earlier in Norway and Sweden than in Denmark which only can be explained by national peculiarities. What I wish to do is to prevent overemphasizing particular causes reserved to one country in order to develop a more "balanced" mid-range theory of comprehensive education in Scandinavia. The overall argument that I wish to put forward—as space allows—is that common to the Scandinavian countries was a specific form of political mobilization and integration that was absent from Britain and the European continent (treating Scandinavia as extra-continental as Nordics are inclined to do) and that paved the way for a radical way of schooling.

Political Liberalism

The most important event in the development of comprehensive education in Scandinavia is properly the introduction of the middle school at the turn of the nineteenth century, as it formed the pivot of a linear school system. Most other European countries did not have a similar good starting point when comprehensive types of schooling were introduced especially after World War II. In Germany, for example, the nine-year secondary school, the *Gymnasium* was maintained intact in parallel to the technical school, the *Realschule*, and the main school, the *Hauptschule*, which constituted such an obstacle when efforts were made to integrate these school types. This problem was simply absent in Scandinavia because the secondary school was turned into an upper secondary school type by abolishing the bottom part of it.

Common to Scandinavia was that the middle school was introduced by Liberal parties who sought to use education as a means of creating "class circulation." The Liberal parties in Scandinavia were formed in opposition to the conservatives during the 1860s to the 1880s and came into power at the turn of the twentieth century, in Norway a couple of decades earlier, which was used successfully to introduce radical reforms on education. Seen in contrast to Germany where liberalism as a political movement failed during the Bismarck era in the 1860s and 1870s and gave way for a conservative dominance mainly expressed by the Catholic Center Party, it makes sense to stress the success of liberalism in Scandinavia.

However, it would be wrong to argue, that liberalism was particularly powerful in Scandinavia, even though the Liberal parties succeeded in obtaining governmental power. On the contrary, as the comparative historian

Luebbert rightly argues, the Liberal parties in any of these Scandinavian countries made a relative weak showing (Luebbert, 1991, p. 58). The Liberal parties gained their strongest foothold in Denmark and were weakest in Norway and especially in Sweden where the party hardly had any significance during the late nineteenth century and was soon bypassed by the Social Democrats. None of the Liberal parties were able to obtain power before the introduction of parliamentary sovereignty, which happened in Norway in 1884, in Denmark in 1901, and latest in Sweden, in 1906, but was only recognized as late as 1917. The reason why liberalism as a political movement was weaker in Norway and Sweden than in Denmark was, according to Luebbert, mainly because the Norwegian and Swedish liberal movement were marked by stronger cultural cleavages than in Denmark, which made it very difficult to function as a united political force. In contrast, the Danish liberal movement was based on a rather homogeneous mass of the rural middle class (including small independent farmers) that made it possible to obtain a more sustained political influence (p. 63). The lines of conflict that debilitated liberalism as a political movement, in all the Scandinavian countries but less so in Denmark, varied, but the most common were between the urban and rural middle classes and between religious communities that originated in the preindustrial period. These preindustrial cleavages were sufficiently strong to prevent an adequate fraction of the middle classes from being unified by liberal mobilization.

In Norway, where the Liberal Party (*Venstre*), which actually had a somewhat homogenous political base forged through a campaign of antiurbanism, antielitism, and anti-Swedishness, obtained governmental power in 1884, but soon thereafter split as this binding force started to crumble. This split was never mended. The first split occurred almost immediately after obtaining governmental power. The religious fundamentalists refused to cooperate with the more broadminded urban radicals and established their own party, the Moderate Liberals, which gained sufficient support to deprive the Liberal Party of a majority in the election of 1889. During the following decades the Liberal Party was so torn apart by conflicts that it was not able to regain its governmental power until the final decade of the nineteenth century. And by that time the party was dominated by more conservative farmers from the western part of the country who subordinated the urban Liberals committed to social reform. The Liberal Party was never able to overcome the cultural cleavages and was consequently excluded from making an influential coalition that combined the urban and the rural middle classes. The result was that after 1884 the Liberal Party alternated government majorities with the Right (*Höyre*) until the World War I (Mjeldheim, 1984).

The political mobilization of liberalism in Sweden was quite similar to that of Norway albeit in a more muted fashion. In neither Sweden nor Denmark did the urban-rural conflict reach the same level of intensity of cultural struggle as it did in Norway. Only in the final decades of the nineteenth century did there appear a Liberal Party in Sweden committed to parliamentary sovereignty and universal suffrage. This party was represented in the Second Chamber and supported by the farmers that constituted the backbone of that chamber. The urban-rural divide that hampered Swedish liberalism was most apparent in the Liberals' failed attempts to establish Second Chamber parliamentary sovereignty and to reform the franchise, a franchise that was in fact as restrictive as in Prussia.[1] The First Chamber, dominated by members from the rural aristocracy, turned down numerous motions on franchise reform as well as on social and education reform put forward by the Second Chamber Liberals who increasingly received mass support especially from the early 1890s. However, the Staff government, a short-lived minority Liberal government from November 1905 to May 1906, was not able to overcome the cleavage between the more reluctant rural wing of the party and the demanding urban radicals, which added to its failure to sustain as an influential party. The Conservative government took over until 1911, and was again followed by a Staff government that once more failed, because it could not overcome this rural-urban split.[2] As a consequence the party was quickly surpassed and marginalized by the Social Democrats. The fast mobilization of the working class is a further reason of why Swedish liberalism passed so quickly (Thulstrup, 1968).

Danish liberalism suffered for more than a half a century by a similar conflict between town and country. The successive advance and relapse of Danish liberalism was closely reflected in its ability to make urban-rural coalitions. In contrast to Norway and Sweden, the Danish Liberal Party (*Venstre*) was able to dominate politics for a longer period, especially in the period of 1901 to 1913. However, soon thereafter the party also started to be eclipsed by Social Democrats.

It was in particular in the party's fight for parliamentary sovereignty and universal suffrage prior to 1901 that the seeds of a split were sown between the party and their urban allies, the National Liberals. This was a bourgeois party, which was dominant during the period of 1848–1864. The National Liberals and the Liberals had successfully worked together in introducing the so-called June Constitution in 1848, but soon thereafter the National Liberals moved to the Right and formed a coalition with the rural aristocracy. Together they controlled the upper house, the *Landsting*. The fight for democracy was then taken up by the Liberals who were now in direct confrontation with the National Liberals through the majority it always had in the lower house, the *Folketing*, after 1872. This struggle that went on until 1901, where the Liberals obtained governmental power, was in fact a major effort to rebuild the urban-rural coalition that had been shattered during the 1850s. The success of a renewed coalition finally provided the sufficient margin to crush the Right in the *Folketing* elections. In the election of 1901, for example, the Right only won 8 out of 102 seats. The *Venstre* and its offshoot the *Radikale Venstre* (Radical Liberals) that was established in 1905, governed until 1920 almost always with the support of the Social Democrats (Salomonsson, 1968). Although liberalism arrived late in Denmark and was made unstable by the division of the party, it was more successful here than in Norway and Sweden. In contrast to Norway and Sweden a successful re-alliance between urban and rural interests in Denmark was reinforced toward the end of the century. In neither Norway nor Sweden was the Liberal Party sufficiently mobilized and integrated politically to sustain its power for a longer period.

The explanation of the successful introduction of the middle school that democratized the education system radically does not, therefore, depend entirely on the strength of liberalism but rather its form. This becomes evident when Britain is taken into account because political liberalism here made its strongest showing in all of Europe and yet this did not lead to any radical education reforms that enhanced social mobility substantively. The explanation of the Scandinavian peculiarity is rather to be found in the origins of class and ideology that constituted the Liberal parties, which sets them apart from all other Liberal parties on the continent and in Britain. The backbone of the Liberal parties was the peasantry. These were not Liberals whose agenda was defined mainly by the urban middle classes as on the continent and in Britain, but movements of independent peasants who sought, above all, to curtail the spending authority and to reject the cultural, linguistic, and religious dominance of cosmopolitan urban bourgeoisie. Regarding education, they argued against the elitist system of the bourgeoisie in order to induce democratic measures to enhance social mobility in the society. The middle school was the means. The urban bourgeoisie had in fact much more in common with continental Liberals than did the peasant Liberals who were shaped by social-liberal values. The ultra liberal views, such as minimum state intervention and self-help, which were expressed in the British Liberal Party and deeply embedded in the Victorian society—as evidenced by the enduring mass popularity of Samuel Smiles's book on Self-Help—was totally unacceptable for most Liberals in the Scandinavian countries.

The successful political mobilization of the peasantry into Liberal parties is a result of the previous organization, initially at the local level, already during the early decades of nineteenth century, and then also at the state level. In Denmark this was made possible due to the early abolition of the adscription in 1788, which allowed the peasants through landownership to improve their social and economic position. In Norway, and especially in Sweden, the mobilization of the farmers went along different lines to that in Denmark, but also here the farmers penetrated governments early in the day (Semmingsen, 1954; Bjørn, 1988; Seip, 1981; Carlsson, 1954). The schooling of the peasantry in public elementary schools, which resulted in the highest literacy rate in Europe, and the popular education received in the so-called Grundtvigian *volk* high schools, integrated the peasantry into societies as active political citizens (Wiborg and Korsgaard, 2006). The making of the peasantry into an independent class that subsequently constituted a liberal power force with social liberal views strong enough to crush the Right is unique to Scandinavia. In no other European country, were conservatives weakened to such an extent as a result of peasant mobilization as they had been in Scandinavia.

The Different Timing of Early Comprehensive Education

This uniqueness is mirrored in the reforms of education that were introduced during the nineteenth century and the inter bellum period at almost all levels. The most radical reform, which introduced the middle school, was promulgated to break down the elite school system, and had no counterpart in other European countries. In continental European countries the eight/nine-year secondary schools were maintained intact, though slowly opening them up to pupils from less fortunate backgrounds. In this context, it is peculiar that the similar-looking Scandinavian countries introduced the middle school at very different times—in Norway in 1869, more than 30 years earlier than in Denmark and Sweden in 1903 and 1905, respectively. This requires a comparative explanation that has to rest on national differences. The reason as to why the middle school was introduced earlier in Norway than in the neighboring countries has mainly to do with the fact that a bicameral political system in the traditional sense was never introduced in Norway. The aristocracy was simply not sufficient in size to form a first chamber.[3] The Norwegian society was, unlike even the Danish and Swedish societies, much more egalitarian as it was the bourgeoisie, not the aristocracy, that formed the upper class. The consequence of this politically was that a conservative resistance from a first chamber could not arise to reform from that side. Had it not been for the absence of a first chamber, the Liberal minister Johan Sveidrup would presumably not have been able to introduce the middle school so soon even with the extensive support he received from fellow Liberals with farming backgrounds. Already in the 1830s, the majority of the government was farmers (Seip, 1981; Dokka, 1967).

The organization of Swedish politics was marked by its history as a great power with centuries of accrued legitimacy behind it. In Norway there was no great power legacy. The state bureaucracy had been subject first to Danish then Swedish power for more than 300 years. Swedish society was also much more stratified than in Norway and even Denmark, containing an influential landed aristocracy wedded to the authority and status of the monarchy and state bureaucracy. In Sweden a bicameral system was introduced in 1866 where the landed aristocracy dominated the first chamber and Liberal farmers the second chamber. During the last two decades of the nineteenth century the Liberal Party put forward petitions numerous times that were consequently turned down by the first chamber. However, the farmers' breakthrough into the parliamentary scene was crucial since they played a major part in the short liberal era in acting as counterweight to the dominance in the first chamber. When they managed to make an entrance to the first chamber at the turn of the century whereby the landed aristocracy gradually was phased out, it was finally possible for the Liberal education minister, Fridtjuv Berg, to introduce a middle school (*realskola*) in 1905 (Sjöstrand, 1965).

In Denmark the middle school could have been introduced already in 1871, since the Liberals by this time formed a majority government. Also the National Liberals had made sufficient concessions to the Liberals, which took the spike out of the opposition. However, the reason why this opportune situation was not fully exploited has mainly to do with a split within the Liberal Party itself. Within the party an influential separatist group of national-romantics (followers of Grundtvig) fought successfully against the party's plan to introduce a middle school.[4] According to the national-romantics, the elementary school, which was seen as expression of the national spirit of the people, was not to be attached to the upper secondary school, because it was deeply embedded in the Humanistic culture of classical languages. Only when the upper secondary school had removed the "Roman yoke" or the "dead languages" and replaced them with a popular curriculum focusing on the mother tongue and national history and poetry, could a middle school be established as a link between the two school types. It was only when the influence of this group of national-romantics on party politics was in decline, that it became possible for the Liberal education minister I.C. Christensen to introduce the middle school in 1903 (Skovgaard-Petersen, 1976).

The Rise of Social Democracy

Immediately after World War I there was a strong interest in the further development of the moderate comprehensive school system in the Scandinavian countries. Even though the middle school became very popular since children from mixed social backgrounds increasingly became enrolled, it was, nevertheless, expressed as a concern that children with less academic ability were excluded from the middle school and instead had to enrol in two extra top classes in the elementary school. This selection, which especially effected children from the rural areas, became the focus of attention of politicians and educators and resulted in the abolishment of the middle school, which paved the way for a seven-year comprehensive school (with streaming in the top two classes) that became common for all. The seven your comprehensive school was introduced in Norway in 1920, in

Sweden in 1927, and in Denmark as late as 1958. After World War II the seven-year comprehensive school was extended even further, from seven to nine years in order to postpone selection even further into the future (Markussen, 2003; Skovgaard-Petersen, 1976; Richardson, 1999).

One important factor of this development the Scandinavian countries held in common was and still is the powerful influence of social democracy. The Social Democratic parties took over to a large extent the education policy program of the Liberal parties and grounded it in socialistic ideology and obtained a long-lasting impact on education policy. However, it is rather peculiar that these preindustrial "peasant societies" would produce such similar and influential Social Democratic parties.

The Danish sociologist Gösta Esping-Andersen argues that the early political mobilization of the peasants steeped in social-liberal values was of great importance to the rise of the Social Democratic parties in the Scandinavian countries (1985). The alliance that the Social Democrats made with the Liberal Party, for example, fighting together for parliamentary and universal suffrage, helped the party to escape from political isolation. The Social Democratic parties in other countries that did not form alliances with other parties, like the German sister party, resulted in isolation and lost power (Hodge, 1994). The alliance between the Social Democrats and the Liberals in the Scandinavian countries could be made, because neither of them was rooted in radical socialism and liberalism respectively. However, there were a few intellectual socialists with radical views in the Social Democratic parties, but they did not dominate the official policy of the parties and accepted quickly to work within the parliamentary framework of democracy.[5] The Liberals learned that "the socialists were not necessarily a threat" and the "socialists discovered that significant strides could be made through class collaboration. It seemed logical that additional reforms were possible through ad hoc alliances" (p. 7).

Furthermore, according to Esping-Andersen the rising power of the Scandinavian Social Democratic parties after World War II relied on the fact that they were able to ally themselves with the white-collar middle class. This was of great importance of their continuing success since the number of workers and, therefore, potential voters in spite of capital intensive industrialism at the time did not increase. The state also played a role in this due to the expanding public sector and it was here the white-collar middle class was concentrated. This class, which had a larger affinity to labor than to the private sector employees, became the backbone of the social democratic voting force. Bourgeois resistance was generally modest and, even under limited suffrage, the Social Democrats had managed to gain representation and effect policy at both the local and the national level. The alliance between the Liberal parties and the Social Democratic parties is a unique feature of Scandinavian politics, which has had a lasting impact on education politics as *consensus* seeking politics. It is very hard to find any piece of legislation that is not a product of this unique political tradition of broad agreements.

Even though the Scandinavian countries developed similar comprehensive school systems, the development of them was marked by the differences in the relative power of the Social Democratic parties in each of the Scandinavian countries. The Social Democratic Party in Sweden had more often been in power than its sister parties in Denmark and Norway. This meant that the Swedish party was much less dependent upon the Liberals. In Denmark and Norway the Social Democratic Party mostly had to be in coalition with the Liberals, although this was only the case for Norway after the 1960s. Hitherto, the Norwegian Social Democratic Party was exceptionally powerful. This pattern of political influence can explain why the comprehensive schooling was introduced earlier in Norway and Sweden than in Denmark. In Denmark the "delay" was caused by a stronger political liberalism in comparison with Norway and Sweden. Especially after World War II it was no longer the "progressive" party in obtaining comprehensive education. On the contrary, the party in general tried now to hamper the Social Democratic policy of comprehensive schooling by maintaining tracking.

Norway

The Social Democratic Party gradually obtained political influence that culminated in 1933 when it came into power. Even though the Social Democratic Party went through some turbulent years—it was a member of the Comintern for a few years—it had in conjunction with the Liberals—a liaison between the Social Democratic Party and the Liberal Party was consolidated in the late inter-bellum period—almost nothing to fear from the Right. Already in 1920 an act that enhanced the principle of comprehensiveness in the education system was introduced. It was decided on a basis of broad political consensus that the duration of the middle school was to be reduced from four to two years so that the duration of the elementary school was seven years, and that the middle schools should only be able to obtain financial support from the state if they enrolled pupils who had graduated from the seven-year public elementary school.[6] In the school act of 1936 the middle school was finally abolished, and now the five-year academic secondary school was in direct continuation of the seven-year elementary school. The less academic pupils were enrolled in a three-year secondary technical school (*realskole*). However, the first two classes of the secondary technical school were integrated in the first two classes of the academic secondary school in order to create a strong link between elementary and secondary education. In the period of 1954 to 1969 the Social Democrats whose power now culminated introduced a nation wide "experiment" with comprehensive education. On the basis of a broad political consensus, a school act in 1969 on the nine-year comprehensive school was finally introduced. However, most counties by now had already introduced comprehensive education. Soon after mixed ability classes were also introduced. In 1994 the comprehensive school was extended from nine to ten years by lowering the school start by one year (Dokka, 1981; Telhaug, 1994).

Sweden

In the ante-bellum period in Sweden, the Social Democratic Party gradually became the largest party. However, it was an unstable time politically, because it was not possible for either a single party or a coalition of parties to form a durable government. Furthermore, the Liberals were not able to mobilize efficiently during these years mainly because the party was a hybrid of temperance movements, intellectuals, and white-collar groups with no strong links to interest groups. The Social Democratic and Liberal coalition formed in 1917 had difficulties in creating a tight connection between elementary school and the upper secondary school. The comprehensive school plan (a seven-year elementary school [*bottanskola*], which the Social Democratic education minister Värner Ryden put forward), collapsed. Furthermore, he was not fully backed up by the Liberals, since they did not anymore fight for this type of schooling. In 1927, a decision was finally made in relation to comprehensive education that was a compromise between the Social Democrats and the now two Liberal parties, the so-called *Frisinnade* and the *Bondeförbundare*. The compromise was called the "double attachment," which implied that, on the one side, a six-year comprehensive school should be followed by a four-year middle school (mostly in small towns) and, on the other side, a four-year comprehensive school should be followed by a five-year middle school (mostly in the cities). This compromise upheld the parallel system that the Social Democrats tried to break down; however, the parallel system was gradually broken down anyway especially due to the abolishment of state financial support to the private preparatory schools of the middle schools.

The Social Democratic Party went into the postwar period just as strong as the Norwegian sister party; however, some stagnation occurred during the 1950s that hindered the party in forming a majority government. In order to maintain power it had, therefore, to establish an alliance with the Liberal Party called the *Bondeförbundet*. The Liberal parties, except for the *Bondeförbundet*, were rather critical of the Social Democratic policy regarding education, which forced the Social Democratic Party to some extent to tone down their ideology in order to work with the Liberals on education legislation. A compromise between the Social Democrats and the Liberals was reached in 1950 where an act on the experimentation of a nine-year comprehensive school was introduced. In the following years, the Liberals could not anymore become a serious hindrance of social democratic policy since the number of mandates was gradually reduced. In 1957 the Social Democrats formed a majority government and in 1962 they introduced a school act that consolidated the nine-year comprehensive school system. However, by now most counties had already introduced this school type. Furthermore, it was decided in 1968 that the streaming (9 tracks) in the top classes of the comprehensive school should be integrated. And in 1980—repeated again in 1994—the setting was also abolished in order to create mixed ability classes (Herrström, 1966; Isling, 1984; Richardson, 1999; Marklund, 1980).

Denmark

In Denmark the Social Democratic Party did not enjoy similar power as the Norwegian and Swedish parties, wherefore it had to cooperate with mainly the Liberals to a larger extent. In spite of increasing power of the Social Democrats during the inter-bellum period it is a rather surprising fact that the middle school was not, as in Norway and Sweden, abolished in order to create a seven-year comprehensive school. Instead a new type of middle school was established in 1937 in parallel to the old middle school of 1903. The aim of the new middle school was to facilitate the needs of pupils who did not have the academic ability to enrol in the academic-oriented middle school. Even though the act was introduced by the education minister Jørgen Jørgensen (*Radikale Venstre*) from a Social Democratic and Social Liberal (*Radikale Venstre*) coalition government, the main thinking behind the act was to facilitate the needs of students by the means of a bifurcated school system. This act shows that it was not all Social Democrats that supported the idea of creating a comprehensive school, but they were the minority.

In the postwar period there was a strong effort to break down this selective middle school system in order to create a unified school system that aimed at equality and social integration. Since the Liberals, in contrast to Norway and Sweden, also at this time played a stronger political role, the introduction of a comprehensive school system took longer. The Liberals accepted in general the idea of comprehensive schooling; however, they didn't accept the social democratic requirement of the long duration of comprehensive schooling (between 7 and 12 years) and the abolishment of streaming and setting.

The middle school was finally abolished in 1958 and a seven-year comprehensive school introduced by a majority government that consisted of the Social Democrats, the Social Liberals and a small party called the *Retsforbundet*. Since the parties could not come to an agreement on the question of selection a compromise had to be met, which implied that selection in the last two classes of the elementary school (grade 6 and 7) should be enforced unless the majority of the parents wanted those to be mixed ability classes. However, it was the possibility of postponing the selection until the end of grade seven that later became the rule, since an increasing number of parents opted for this. The seven-year comprehensive school was now gradually introduced nation wide.

During the 1960s and 1970s the Social Democrats radicalized their program by demanding a nine- sometimes twelve-year comprehensive school, mixed ability classes, and the abolishment of exams and grades entirely. The Liberal opposition to the Social Democrats grew even stronger since they argued for the selection in order to enhance academic standards. This battle was finally solved through a compromise across parties from Left and Right, when the Social Democratic education minister, Ritt Bjerregård, introduced the nine-year comprehensive school in 1975. In order to arrive at a compromise the parties had to give up some of their leading issues. For example, the Social Democrats had to accept that exams and grades had to be maintained as well as tracking had to be implemented at grades eight and nine. However, Ritt Bjerregård was able to negotiate that tracking at grades eight and nine could be abolished after approval of the local school authorities. After a change of government in 1991 the education minister, Ole Vig Jensen (*Radikale Venstre*), introduced mixed ability classes in a school act of 1993 (Bregnsbo, 1971; Markussen, 2003).

Concluding Remarks

The overall comparative explanation of why a fully comprehensive school system could be introduced in Scandinavia has mainly to do with the unique tradition of consensus seeking politics between Left and Right. The making of the peasantry into an independent class that subsequently constituted the Liberal Party with socially liberal views strong enough to crush the Right, and the rise of Social Democracy who welded an alliance with the Liberals goes far in explaining how a radical tradition of education could be introduced through broad coalitions. Whether this unique tradition in Scandinavia will survive is another matter. A neoliberal turn in education politics during the last ten years or so may undermine the principle of comprehensive education in the future. Academic standards have been enhanced as a result of the Program for International Student Assessment (PISA) studies and the Organization for Economic Cooperation and Development (OECD) reports. Yet, the introduction of testing and parental school choice is still in its infancy and, therefore, it is too early to say anything conclusive about its effect on comprehensive education. There is no doubt that it will have an effect, but to what extent only the future can tell.

Notes

1. As late as 1896, only 6 percent of the population (about 20 percent of the adult males) had the right to vote.
2. This time the conflict was the question of military spending that the urban radical wished to curtail in favor of social reform.
3. There were about three aristocratic families of Danish/German origin.
4. The national-romantics were inspired by the priest and poet N.F.S. Grundtvig who shared in many respects the national romantic ideas of the German J.G. Herder.
5. The Norwegian Social Democratic Party was an exception to this from 1919 to 1927 since it was member of Comintern.
6. The act was supported by the Liberal Party, the Social Democratic Party and the Labour Democrats (*Arbeider-demokraterne*).

References

Bjørn, C. (1988). *Det danske landbrugs historie III: 1810–1914*. Odense: Landbohistorisk Selskab.

Bregnsbo, H. (1971). *Kampen om skolelovene af 1958: En studie i interesseorganisationers politiske aktiviteter*. Odense: Odense University Press.

Carlsson, S. (1954). Ståndsupplösning och demokratisering i Sverige efter 1780. In *Ståndssamhällets upplösning i Norden*. Åbo: Turku.

Dokka, H-J. (1967). *Fra allmueskole til folkeskole*. Bergen: Universitetsforlaget.

———. (1981). *Reformarbeid i norsk skole. 1950-årene—1980*. Oslo: NKS-Forlaget.

Esping-Andersen, G. (1985). *Politics against market*. Princeton, NJ: Princeton University Press.

Herrström, G. (1966). *1927 års skolreform: En studie i svensk skolpolitik 1918–1927*. Stockholm: Svenska Bokförlaget.

Hodge, C.C. (1994). *The trammels of tradition. Social democracy in Britain, France and Germany*. Westport, CT: Greenwood Press.

Isling, Å. (1984). *Kampen för och mot en demokratisk skola 1*. Stockholm: Sober Förlags AB.

Jensen, S. (1954). Stands—og klasseforhold i Danmark i tiden mellem slutningen af 1700-tallet og i dag. *Ståndssamhällets upplösning i Norden*. Åbo: Turku.

Leschinsky, A. and Mayer, K. (Eds.) (1999). *The comprehensive school experiment revisited: Evidence from Western Europe*. Berlin: Peter Lang.

Luebbert, G.M. (1991). *Liberalism, fascism, or social democracy*. New York: Oxford University Press.

Marklund, S. (1980). *Skolsverige 1950–1975. 1950 års reformbeslut, Del 1. 1950 års reformbeslut*. Stockholm: Liber UtbildningsFörlaget.

Markussen, I. (2003). Dannelsessyn og drivkræfter bag enhedskolens fremvækst. In Slagstad (ed.), *Dannelsens forvandlinger*. Oslo: Pax Forlag.

Mjeldheim, L. (1984). *Folkerørsla som vart parti. Venstre frå 1880åra til 1905*. Oslo: Universitetsforlaget.
Richardson, G. (1999). *Svensk utbildningshistoria. Skola och samhälle förr och nu*. Lund: Studentlitteratur.
Salomonsson, E.S. (1968). *Den politiske magtkamp 1866–1901*. København: Jørgen Paludans Forlag.
Seip, J.A. (1981). *Utsikt over Norges historie, Bind 2*. Oslo: Gyldendal.
Semmingsen, I. (1954). Standssamfunnets opplösning i Norge. i: *Ståndssamhällets upplösning i Norden*. Åbo: Turku.
Sjöstrand, W. (1965). *Pedagogikens Historia 3:2*. Lund: CWK Gleerups Förlag.
Skovgaard-Petersen, V. (1976). *Dannelse og demokrati*. København: Gyldendals pædagogiske bibliotek.
Telhaug, A.O. (1994). *Norsk Skoleutvikling etter 1945*. Oslo: Didakta Norsk Forlag.
Thulstrup, Å. (1968). *Svensk Politik 1905—1939*. Stockholm: Bonnier.
Wiborg, S. (2004). Education and social integration: A comparative study of the comprehensive school system in Scandinavia. *London Review of Education* 2 (2).
Wiborg, S. (with Korsgaard, O.) (2006). Grundtvig—The key to Danish education? *Scandinavian Journal of Educational Research* 50 (3) (July): 361–82.

Chapter 9

Missing, Presumed Dead? What Happened to the Comprehensive School in England and Wales?

David Crook

Introduction

In 2004, a double-page *Daily Mail* headline, above three barely related education articles, one by the arch critic of state education, Melanie Phillips, screamed "Death of the comprehensive"? (*Daily Mail*, July 9, 2004, pp. 6–7). The placement of the question mark is revealing, but also contentious. The immediate suggestion is that an entity that lived and breathed *may* now be dead. But how can we be sure? Without question, something of note has happened. There is no specialist schools program in Wales, but across the length and breadth of England both words forming the term "comprehensive school," which would once have featured on end-of-driveway entrance signs, have been vanishing. Today, the former "Bash Street Comprehensive School" is more likely to be styled "Bash Street College of Technology" (or perhaps Arts, Engineering, Languages or Sport). It may even be the "Bash Street Academy."

The title of this book, *The Death of the Comprehensive High School*, tempts its contributors to apply the analogies of sickness, health, life, death, survival, and extinction. Forty years ago, there was a widespread expectation that, long before 2007, historians would have borne witness to the death of selective secondary schools, rather than considering, as we now are, the suspected death of the comprehensive. In his speech to the 2002 Labour Party conference, Tony Blair suggested that the comprehensive's days were over. "We need to move to the post-comprehensive era," he declared, "where schools keep the comprehensive principle of equality of opportunity but where we open up the system to new and different ways of education, built round the needs of the individual child" (*The Times*, October 2, 2002, p. 10). Though no death certificate or body to bury has emerged, the orderliness of historical periodization may now demand that "the comprehensive era" should take its place alongside "the Revised Code era," "the school board era" and "the era of child-centeredness." Some steps in this direction have already been taken (*Observer*, January 7, 1996, p. 18; Haydn, 2004).

The remainder of this chapter is divided into five sections. The next part charts the history of English and Welsh comprehensive schooling, providing some case notes relating to the sickness and health of this elusive patient. The third section examines truth and bias, and, drawing upon personal experiences of selective schooling, it underlines the sometimes-overlooked point that comprehensives were developed as an alternative. Next, the chapter considers why the elasticity of the term "comprehensive school" once thought to be an advantage, ultimately proved harmful. This is followed by a section that considers the successes and failures of comprehensive schooling, and, finally, a brief conclusion.

English and Welsh Comprehensive Schooling, 1925–2007

The comprehensive (or multilateral) school movement in England and Wales may be traced back to a 1925 fact-finding tour of nonselective high schools in the United States and Canada by a young Board of Education civil servant, Graham Savage. Savage was to become Education Officer for London in 1940, where demands for universal and free secondary schooling became conflated with calls from some Socialists for a common school. But the climate for radicalism proved bleak, particularly when, despite finding the multilateral an "interesting and attractive" idea, the 1938 Spens Report endorsed a system of segregated grammar, modern and technical secondary schools, underpinned by "11-plus" psychometric tests (Board of Education, 1938, p. 291).

Neither the 1943 White Paper, *Educational Reconstruction*, nor the 1944 Education Act for England and Wales wholly endorsed the Spens line. The former stated that "There is nothing to be said in favour of a system which submits children to a competitive examination on which, not only their future schooling but their future careers may depend" (Board of Education, 1943, para. 17, p. 6), while in requiring that local education authorities

(LEAs) "shall afford for all pupils opportunities for education offering such variety of instruction and training as may be desirable in view of their different ages, abilities, and aptitudes" (Education Act, 1944, section 8, p. 4), the Act did not proscribe different types of experience being offered in a single secondary school. When Clement Attlee's Labour Party won a landslide victory in the 1945 general election, supporters of the common school were hopeful that central government would encourage local interest in nonselective schooling. In fact, the opposite was so. Attlee's minister of education, "Red" Ellen Wilkinson, embraced the 11-plus in the name of working-class opportunity and conspicuously failed to endorse the comprehensive school (Kerckhoff et al., 1996, pp. 18–19).

The postwar years saw several large Labour-controlled cities, most notably London (London County Council, 1947), unveil radical blueprints for comprehensive schooling. The West Riding of Yorkshire, under Conservative leadership, also pressed for multilateral schools at this point, but the ambitions of such localities were stunted by obstruction from within the Ministry of Education and Inspectorate. Instead, "judicious experiments" with nonselection only were sanctioned, with the Welsh island of Anglesey leading the way. It was here, not in London, Manchester or any other large city, that the first "true" comprehensive school was formed in 1949, by uniting the grammar and secondary modern schools on opposite sides of a road.

Only slowly, during their long years of political opposition, between 1951 and 1964, did the Labour Party warm to comprehensive schools. Many leading Socialist figures on the national and local stages had, like Wilkinson, been beneficiaries of the 11-plus and their instincts were to protect the "scholarship ladder" and local grammar schools. Other party members fretted about the widely held understanding that a comprehensive school needed at least 1,500 children in order to maintain viable group sizes for public examinations. In 1955, Labour's Earl of Listowel sought to reassure his fellow peers by referring to a preliminary survey of 14 comprehensives, the largest of which had only 1,254 pupils on the roll. Such schools needed to be nurtured, he maintained: they would not be given a fair chance "if in every case the neighbouring grammar school decides to stand out" (Hansard, House of Lords, Vol. 190, Cols. 1111, 1114, February 9, 1955). This was a pointed reference to the actions of the Conservative education minister Florence Horsbrugh, who, in the previous year, had intervened to prevent closure of Eltham Hill Girls' School, one of five schools—and the only grammar—scheduled by the London County Council to make way for the capital's first purpose-built comprehensive (Simon, 1991, pp. 171–72).

In June 1958, three years after succeeding Attlee as Labour leader, Hugh Gaitskell launched a pamphlet, *Learning to Live* (Labour Party, 1958), which commended comprehensive schools. Anticipating by some seven years Tony Crosland's Circular 10/65, Gaitskell proposed to "ask local authorities to submit plans to abolish the permanent segregation of children into different types of schools at 11," but the force of this policy was reduced by the assurance that "we shall leave them plenty of latitude as to the way in which, and the speed at which, they do this" (*The Times*, July 7, 1958, p. 5). Some 46 "experimental" comprehensive schools were operating by this time, but very few approached the profile of an "authentic" comprehensive, housing 25 percent of "grammar-school types." Virtually none were products of Anglesey-style amalgamations: some had begun life in relinquished school premises, while others were new builds, serving urban working-class housing estates on sites devastated by Luftwaffe bombing.

Aside from the Kidbrooke controversy of 1954–1955, the emergence of early comprehensive schools barely threatened the position of grammar schools. In *The Essential Grammar School*, published in 1956, the Headmaster of Watford Grammar and distinguished wartime hero, Harry Rée, later to experience a road-to-Damascus conversion, was dismissive. Revealing his absolute confidence in the science of psychometrics, he wrote:

> Already in some good Secondary Modern Schools children of average ability are able to enjoy a full and valuable education, because they cannot be disheartened by the proximity of very clever children. They can achieve distinction and a position of responsibility among their peers. The chances of this happening in a Comprehensive School are reduced to almost nil by the presence of the faster-thinking Grammar School stream, for the latter not only have the advantage in direct competition, but also, because they stay on at school for longer (in order to go to the university or to take public examinations), it is they who fill the positions of responsibility, who have the honour of representing the school in the first teams, in the school orchestra, in debates, and so forth. (Rée, 1956, pp. 19–20)

In 1959, the Conservative education minister, Lord Hailsham, attributed to the Labour opposition a plan to "kill the grammar schools" (*The Times*, September 28, 1959, p. 7), but it would be more accurate to say, at this point, that dissatisfaction with the 11-plus exceeded enthusiasm for a national system of comprehensive schools. Evidence from academic studies (e.g., by Vernon, 1957; Yates and Pidgeon, 1957) had questioned the rationale for, and methods of, psychometric testing. Too many children with latent abilities, it was suggested, were being misallocated to secondary modern schools, while a disproportionate number of those winning grammar-school places happened also to be "middle class, wealthy or culturally well-endowed" (Benn, 1992, p. 145).

Simultaneously, what the late Brian Simon termed an LEA-led "break out in secondary education" was occurring, "reflecting increased aspirations and mounting frustration on the part of parents . . . and the local politicians representing them" (Simon, 1991, p. 271). It was this significant and rapid grassroots shift that placed comprehensive schooling firmly on the national political agenda by the time Harold Wilson's Labour Party formed a new government, with a majority of just four parliamentary seats, in October 1964. The secretary of state for Education and Science Michael Stewart reported that 68 LEAs were implementing plans or had developed concrete reorganization proposals. A further 21 authorities were said to be contemplating going comprehensive, leaving only 59 that were not (Hansard, House of Commons, Vol. 702, Col. 1784, November 27, 1964). At the beginning of the following year Stewart announced that LEAs would shortly be asked to submit plans for the reorganization of their secondary schools on comprehensive lines. He did not preclude the possibility of legislation, but anticipated "that we shall have a highly cooperative response from local education authorities and I am proceeding at present in that hope" (Hansard, House of Commons, Vol. 705, Col. 391, January 21, 1965). As a further signal of how quickly the secondary school landscape had changed in less than a decade, it is interesting to note that Harry Rée, the former champion of grammar schools, was, by this time, a leading author of reorganization plans for the city of York, where he had become a university professor four years earlier (*The Times*, January 8, 1965, p. 13).

It fell to Tony Crosland, Stewart's successor, to issue the long-anticipated Circular 10/65 (Department of Education and Science, 1965) on July 14, 1965. Six alternative methods of "going comprehensive" were identified in this non-statutory document, with the 11–18 "all-through" solution, Crosland's own preference, heading the list of reorganization models. Three alternative two-tier schemes were identified, each contemplating secondary education beginning at 11 but with transfer to another school (or sixth-form college) at ages 13, 14, or 16. Authorities were also encouraged to consider whether three-tier schemes, involving 8–12 or 9–13 middle schools, would help to effect reorganization. It was an extensive menu designed to tempt those LEAs that had not already embarked upon reform, but one that, in retrospect, added to the identity crisis of the comprehensive school.

The controversy caused by the Circular manifested itself in heated local and national debates and in divisions between ministers and civil servants at the Department of Education and Science (Dean, 1998). Late one night in 1965, according to his widow, Susan, Tony Crosland came home from the House of Commons in a filthy mood after a day battling with his officials to declare "If it's the last thing I do, I am going to destroy every ****ing grammar school in England. And Wales. And Northern Ireland" (Crosland, 1982, p. 148; Interview with Susan Crosland, *Comp*, Programme Two, BBC Radio 4, September 8, 2005). It is impossible to estimate the extent to which this story, which is frequently (mis)represented as evidence of the Labour Party's disdain for grammar schools, did damage to the comprehensive movement. Certainly, it has harmed the political memory of Crosland, who died suddenly in 1977. Far from being hostile to excellence in education, he was an intellectual, anxious to win the hearts and minds of parents and teachers during the 1966 general election campaign, during which he described the 11-plus eleven-plus as "an absolute curse to children in this country," "a chancy business," and "unjust" (March 1966 election broadcast footage, *The Schools Lottery*, Programme One, BBC Two, March 27, 2006).

With a new majority of 97, Harold Wilson's position seemed much stronger after his second general election victory, but his party's policy toward comprehensives rested upon the cooperation of LEAs. Some were already engaged in comprehensive planning, and many others moved at this point to set up working parties and consultation meetings. For some, though, the inadequacies of existing building stock and finances stymied efforts to introduce change, while a few declared themselves satisfied with existing selective arrangements. Up to this point, comprehensive schooling had not been the party political issue that it might have been, but the belief of Crosland and his successors, Patrick Gordon Walker and Ted Short, that too many Conservative-controlled LEAs were ignoring the Circular, was given substance by the withdrawal of several previously submitted reorganization plans in the wake of sweeping Conservative local election victories in 1967 (Kerckhoff et al., 1996, pp. 32–34). The perceived resistance of around 20, out of 146, LEAs prompted Short to draw up a parliamentary Bill in 1969. Although this became a casualty of Harold Wilson's decision to call the general election that saw Edward Heath's Conservative Party return to power, it signaled the beginnings of a new era of centralization in education (Crook, 2002, pp. 252–53).

Encouraged by the publication of the first two antiprogressive "Black Papers" (Cox and Dyson, 1969a,b), Margaret Thatcher, the new secretary of state, immediately published a replacement circular, Circular 10/70, stating that "Authorities will now be freer to determine the shape of secondary provision in their areas" (Department of Education and Science, 1970). A number of local councils withdrew their plans in order to reconsider, though most decided subsequently to proceed. Thus it was that, in her four-year tenure as education secretary, although Thatcher intervened to "save" some 94 grammar schools identified for closure or redesignation, she found it impossible to halt what she later called the "universal comprehensive thing" and a "great rollercoaster of an idea" (quoted in Chitty, 1989, pp. 54–55). To the chagrin of her followers, she presided over the creation of more comprehensives than any previous or subsequent education secretary.

By the time Labour's Harold Wilson became prime minister again, at the end of February 1974, there were more than 2,000 comprehensive schools in England and Wales, attended by around 60 percent of secondary-age children. Many lacked a balanced intake of pupils, however. A small number had been colonized by the middle classes, but many more comprehensives were secondary moderns in disguise. A new circular, Circular 4/74 (Department of Education and Science, 1974), was issued, with the promise of legislation to follow if the reluctant authorities did not now draw up comprehensive school plans. But with a parliamentary majority of just four seats, rising to ten after a further general election in the autumn, Labour was a toothless tiger. In the period 1974–1979, Wilson and his successor from April 1976, James Callaghan, were defeated an astonishing 42 times in the Commons and 347 times in the House of Lords (Norton, 1985, p. 14). It is unsurprising, therefore, that successive secretaries of state were unable during this period to successfully advance educational reforms. The "rollercoaster" ride had ended and the installation, in February 1975, of Margaret Thatcher as leader of the Conservative opposition prompted a pamphlet (St. John Stevas and Brittan, 1975) that actively encouraged "recalcitrant" authorities, including Buckinghamshire, Kent, and Trafford, to defy national policy and preserve their grammar schools.

Legislation duly followed, but the 1976 Education Act became law one month after prime minister Callaghan had cast doubt on the content and quality of British education, and, implicitly, of comprehensives, in a speech at Ruskin College, Oxford. Talk among comprehensive school enthusiasts of "one last push" could not disguise the fact that the project was now in crisis: a recent Court of Appeal ruling against education secretary Fred Mulley had permitted Conservative-controlled Tameside LEA to withdraw previously submitted reorganization proposals and the new secretary of state Shirley Williams was inviting a plurality of understandings of comprehensive education to be framed at a two-day conference. The conference was to confirm Labour's unwillingness, or inability, to differentiate between genuinely comprehensive and quasi-comprehensive solutions adopted by LEAs. "Whatever the geographical area of the pattern of organisation, all have their merits" wrote the under-secretary of state in a lame conclusion (Jackson, 1978, pp. 13–14; Kerckhoff et al., 1996, pp. 40–41).

The exhaustion that resulted from fighting for or resisting, comprehensive schooling during the 1960s and early 1970s determined that neither the election campaign preceding, nor the years immediately following, Margaret Thatcher's general election victory of May 1979, were dominated by calls to bring back grammar schools. Back in 1978, following interviews with Conservative shadow education ministers Rhodes Boyson and Norman St. John-Stevas, the Right-leaning *Times* columnist, Ronald Butt, anticipated a new kind of secondary education landscape, with no wholesale return to the eleven-plus, but smaller comprehensives and a mixed economy of secondary schools, including ones having curricular specialisms (Butt, 1978a,b). It took some time to happen, but this was to be a very accurate prediction of developments following the opening of the first city technology college in 1987.

The commitment to restore "a grammar school in every town," made by Conservative prime minister John Major in 1996 and echoed by several subsequent Conservative Party leaders (*The Times*, March 11, 1996, p. 1; *Evening Standard*, February 14, 2001, p. 2; *Daily Mail*, June 17, 2004, p. 19), proved less attractive to the electorate than to newspaper columnists. Under the banners of "choice" and "diversity," Conservative and Labour governments of the 1990s and early twenty-first century have focused on the evolution of secondary schools, rather than pursuing another bloody revolution. In 1995, David Blunkett, then Labour's education spokesman, promised "no selection, either by examination or interview, under a Labour government" (*Guardian*, October 5, 1995, p. 8), but it emerged subsequently that he had meant to say "no *further* selection." Upon gaining office, Labour's 1998 School Standards and Framework Act made provision for local parents to decide the future of the remaining grammar schools, but the high number of signatures required to trigger a ballot has meant that only one—demonstrating support for the continuation of selection in Ripon—has taken place. Of the new types of state secondary school to emerge in recent years, grant-maintained schools, then foundation schools and academies, were permitted to operate outside LEA control, while soon-to-be-introduced trust schools will reside within LEA funding structures but may be run by such bodies as charities, universities, or community groups. The most spectacular development of recent years, however, has been the proliferation of specialist secondary schools that were once, genuinely or by aspiration, comprehensives. By the end of 2006, some 2,602 English secondary schools, more than 80 percent of the total, had a specialist designation. The new titles for these schools (or colleges, as many have styled themselves) offer a stronger and more popular market brand than that achieved by the "bog-standard" comprehensive, to use the inelegant term coined by the prime minister's former spokesman (*The Times*, February 13, 2001, p. 1). But the discourses of choice and diversity, in combination with new contexts of school accountability and performativity, have created an increasingly-complex secondary school transfer system in which parental preferences are frequently not met. Consumerist language about parents choosing schools masks the reality that oversubscribed state secondary schools in England select some or all of their intake, whether on the grounds of pupil residency in a catchment area, of siblings already attending the school, of religious conviction, of pupil ability (for entry to surviving grammar schools) or of aptitude (in the case of schools having a designated specialism in languages, the performing arts, the visual arts, sport, design and technology, and information technology).

In their anxiety to avoid underperforming or "failing" schools, demand for places in popular state schools frequently exceeds supply by a multiple of ten. Parents "in the know" employ private tutors to prepare their children for entrance tests, properties are purchased or rented on the "right" side of catchment boundaries and some become religiously devout until a testimonial is secured from the local priest. When applications and appeals are unsuccessful, tears are shed, just as they were shed over letters communicating 11-plus results. Today's postcode lottery, where parents and children emerge as winners and losers in the quest for a "good" state secondary school, may barely be regarded as an advance on the former system of selection by 11-plus.

Comprehensives, Histories, Facts, and Distortions

In times of postmodernity, every interpretation of the past might be represented as equally valid, but the tendency for accounts of comprehensive education to serve political ends and to underpin ideological stances has poorly served the cause of historical truth. In one notorious example, published in 1996, A.N. Wilson, the British historian and social commentator, misidentified Shirley Williams as the Labour Party's "architect of the comprehensive system that eventually spelled the death of the vast majority of grammar schools." "With Stalinist thoroughness," Wilson wrote, Williams "waged war on local councils who tried to save the grammar schools," while "teachers unhappy with the half-baked egalitarian notions of mixed-ability classes were discarded." Thanks to her, he concluded,

> Instead of 17.5 per cent of the population being allowed to better themselves by a good education, 100 per cent were forced to lower themselves by a lousy education. The result of the experiment is that England is now the most boorish, ill-educated, stupid nation in Europe. (Wilson, 1996)

In one respect, only, Wilson was overmodest: academic studies of the post-1945 period point to a proportion of between a fifth and a quarter of children—higher in Wales than in England—who proceeded from their local primary to a grammar school. Otherwise, Wilson's account seems outrageously distorted in suggesting that crusaders for comprehensive schooling were motivated by the destruction of educational traditions and excellence.

Another imagined past was offered by Tony Blair in Labour's 2005 White Paper, *Higher Standards, Better Schools for All* (Department for Education and Skills, 2005, p. 1). In contrast to "official past" examples to which McCulloch (2000) refers, the historical element in this foreword barely extends beyond one page, somewhat briefer than the textual analyses it merits from historians of education:

> Pressure, initially from middle-class parents angry with standards in secondary moderns, led to comprehensive schools and the conversion of grammars and secondary moderns in the 1960s and 1970s. But their introduction was often accompanied by all-ability classes, which made setting by subject ability too rare. Many retained their old secondary modern intake, and failed to improve. There were simply not enough pressures in the system to raise standards. (DFES, 2005, p. 1)

This paragraph presents a novel explanation for the birth of British comprehensive schooling and a highly contentious explanation for its decline. There is no mention here of the role of local authorities in developing comprehensive schools. And were comprehensives invented in the 1960s, as might be inferred here? No. Did comprehensive schools promote mixed-ability teaching and reject setting? In some instances, yes, but not everywhere. Were middle-class parents angry with standards in secondary moderns? Probably, yes, but they were angrier still with the 11-plus test that had allocated their children to these schools in the first place.

It is the silence about the 11-plus in the accounts by Wilson and Blair that seems most curious. Even though it is still in use only in areas served by England's remaining grammar schools, of which there are 164, most British adults seem to know someone with a personal experience of the test. In recent newspaper interviews, deputy prime minister John Prescott has painfully shared the story of failing the 11-plus in 1948, of the promised bicycle in the event of examination success never materializing, and of how his girlfriend of the time, who had won a grammar-school place, dumped him by returning a love letter with the spelling mistakes corrected (*Sunday Telegraph*, December 18, 2005, p. 4; *The Times*, April 27, 2006, p. 6).

Autobiographical writings, "blogs," radio phone-ins, and other broadcasts are fast emerging as fascinating social history sources documenting the anxieties that the 11-plus test brought to generations of children and their parents. In a recent television program, Trevor Bayliss, the London-born inventor of the clockwork radio, remembered his parents crying when they learned that he had failed to win a grammar-school place in 1947. The cruelty of selection and notification methods was no less evident a decade later: Greg Dyke, who rose to become director general of the British Broadcasting Corporation (BBC), recalled his brother running down the street shouting "I've failed" and his own experience of knowing that he had passed from the size of the envelope handed to him by the postman. Dyke's contemporary, Nick Ross, another broadcaster, was also fortunate to be selected in 1958, but reflected:

> I can tell you that there was a huge sense of exhilaration and relief, which was only marred when I discovered, within a few days, those friends of mine who hadn't [passed]. . . and, in some cases, it was unfathomable. . . . He was brighter than me. . . . And I've got into grammar school. I've never quite understood that, and even though I was on the winning end of this . . . right from those early days, long before this became a political issue, to a ten-year-old, there was a sense of injustice.

Another person of similar vintage, Sandie Shaw, the shoeless pop siren of the 1960s, recounted that

> the day that I sat my 11-plus I came out of school and my godfather was standing at the gates. He never used to pick me up from school. I used to walk home on my own. And I said to him "What's the matter? What's going on?" And he said "Sandra, I have to tell you, your grandfather died two days ago and Mummy didn't want anybody to tell you because your 11-plus was

so important." (Interviews, *The Schools Lottery*, Programme One, BBC Two, March 27, 2006)

Declan McManus failed the 11-plus in 1965 and proceeded to Hounslow Secondary Modern School in West London. Twenty-five years later, in his stage guise as Elvis Costello, he released a bittersweet song, "Secondary modern," the lyrics of which addressed the sadness of taking "second place in the human race" (Costello, 1980). For Evan Davis, now the BBC's Economics Editor, hatred of the 11-plus was a shared generational experience between himself and his parents (Davis, 2005). Such recollections point to a different "story" of comprehensive education, one that is more satisfactorily captured by Barry Sheerman, currently chairman of the parliamentary Education and Skills Committee:

> We should all remember with some humility that comprehensive education was a grass-roots movement. It was a passion of people from many parts ... who hated the 11-plus, and the social division on which it was based, so much that they and local authorities throughout the country—some of them not Labour—changed the situation. National politicians joined in a bit later. (Hansard, House of Commons, Vol. 443, Col. 1497, March 15, 2006)

One dimension is still missing from this account, however: acknowledgment of the multiple understandings of "comprehensive school" that proliferated from the 1950s.

The Multiple Meanings of the Comprehensive School

From its beginnings, in the 1920s, the term "comprehensive" has been stretched in all manners of directions. Early understandings emphasized social and political aims directed toward creating a more egalitarian society, a more cohesive neighborhood, and a common culture. Writing in 1958, Trevor Lovett, head teacher of the first genuine British comprehensive, in Anglesey, considered that a true example "should be the only school where education usually associated with the secondary phase is provided," with pupils comprising "a fair representation of the normal society in which they must one day play their part" (Lovett, 1958, p. 48).

This idea of the neighborhood comprehensive school always remained close to the heart of Brian Simon, the pro-comprehensive activist and leading British historian of education. In 1994, at the height of a controversy following Tony Blair's decision to send his eldest son to the London Oratory School, some eight miles away from Downing Street, Simon defined comprehensives as "schools recruiting all, or nearly all, children living in the locality" (*Observer*, December 4, 1994, p. 23). But the local variables of urban education, including proximity to selective independent schools and, until the mid-1970s, direct-grant grammar schools, compromised the neighborhood principle long before the slogans of "choice" and "diversity" empowered parents like the Blairs. A 1968 feature in *The Times* focused on Withernsea High School in the East Riding of Yorkshire: as the sole secondary school serving an area of 100 square miles; this was, by default, a neighborhood comprehensive (*The Times*, August 5, 1968, p. 2). But two years later, the head teacher of the futuristic Pimlico Comprehensive School, built near London's Victoria Station, commented that his pupils were drawn from all parts of the city, adding "in no sense are we a neighbourhood school" (*The Times*, October 20, 1970, p. 4).

For several decades it was a common understanding that comprehensive schools needed to be large, having at least 1,500 pupils, in order to maintain curricular breadth and to sustain a viable "sixth form" beyond its provision for 11- to 16-year-olds (see Ministry of Education, 1947). Being significantly larger than the typical population of grammar or secondary modern schools, this was an instant deterrent in some areas, sometimes because split- or multisite comprehensives were an anathema, sometimes because structures for leadership could not be envisaged, but also because of fears about children becoming "lost" in such a vast organization. Declared opponents of comprehensives mischievously played upon these fears: in 1956 Harry Rée cited the view of his Manchester Grammar School counterpart, Eric James, that a comprehensive school required 5,000 on the roll "in order to provide a sufficient number of pupils in the Grammar School stream to offer each other effective competition at a high level, and to make economic a wide diversity of choices for sixth form specialists" (Rée, 1956, p. 19).

Within many comprehensive-minded LEAs, 11–18 "all-through" schools were seen as essential. The 1947 *London School Plan*, which proposed a city-wide reorganization along these lines, became almost a biblical text for two generations of unrealistically optimistic Socialist politicians and Education Officers facing massive organizational and demographic impediments to change. In London and elsewhere, a more flexible approach to structural planning, redesignating grammars as upper-tier schools or sixth-form colleges, might simultaneously have accelerated the pace of change and won over the middle classes, whose flight to the independent sector and to suburban areas having grammar schools or comprehensives that resembled grammars, has left a residue of struggling urban schools, which one former education secretary has said she "wouldn't touch with a bargepole" (*Guardian*, June 25, 2002, p. 1).

It was the bogy of comprehensive school size, as well as dissatisfaction with the 11-plus, that prompted Leicestershire to implement an ingenious workaround after 1957. The "Leicestershire Plan" involved all primary school children moving to a common "junior high" at age 11. Three years later, on the basis of parental wishes and teacher recommendations, some proceeded to a grammar school, the "senior high" in this three-tier system, while the remainder completed a final year in the junior high before leaving school. Although this scheme finds a place in most histories of comprehensive schooling, the Leicestershire model primarily reflected dissatisfaction with selection at 11, postponing this until age 14. It permitted grammar schools to keep their names and stopped short of fully embracing the comprehensive school ideology. In other parts of England, particularly in Conservative shire counties including Bedfordshire, Northamptonshire, and Northumberland, three-tier reorganizations, featuring middle schools, facilitated a positive response to Circular 10/65 and hastened relatively uncontroversial secondary reorganizations.

The policy shifts and inconsistencies of both main political parties added to the comprehensive school's identity crisis. Although frequently presented as a Socialist project, comprehensive schools attracted many local Conservative politicians and some prominent national figures, too. In his cabinets of 1990–1997, John Major was served in the Education Department by a secretary of state Gillian Shephard and a junior minister, Robin Squire, who, in Norfolk and Sutton, respectively, had both campaigned for comprehensive reorganization 20 years earlier (*Education*, November 10, 1995, p. 3; *Observer*, May 9, 1999, p. 5). In the late 1960s, Conservative shadow education minister Edward Boyle found his open-minded views about comprehensives increasingly at odds with those of party hawks (Crook, 1993) and, in opposing Labour's 1976 Bill, Norman St. John Stevas, Margaret Thatcher's frontbench Education spokesman, made clear that the objection was to compulsion, not to comprehensive schools. Straying off message, somewhat, Stevas acknowledged that Conservative councils had helped to pioneer comprehensives and maintained that "The Conservative Party is the true friend of the comprehensive school because we approach this problem practically and not dogmatically" (Hansard, House of Commons, Vol. 919, Col. 230, November 9, 1976).

But the Labour Party's relationship with comprehensive education has been still more vexing, causing public and private lives to regularly and spectacularly collide. Even as Circular 10/65 was being drafted, the maverick C.P. Snow, speaking from the government frontbench of the House of Lords, stated that "comprehensive" was "a rather absurd title" and then conceded that he had sent his own son to Eton College (Hansard, House of Lords, Vol. 263, Col. 161, February 10, 1965). Some lessons were learned from this episode, with London's Holland Park Comprehensive School establishing a reputation for educating the children of Labour cabinet ministers, including Tony Crosland's step-daughters, but the case of Master Snow was neither the first nor the last example of private decisions or unfortunate jibes undermining the party's public principles.

Labour were long haunted by the view attributed to Harold Wilson, whose own children were educated in private schools, that comprehensives could be "grammar schools for all" and that the grammars would be abolished "over my dead body." In fact, Hugh Gaitskell, Wilson's predecessor, had earlier confirmed that the party was seeking "a grammar education for all" and "we want to see grammar school standards—in the sense of higher quality education—extended far more generally" (*The Times*, July 7, 1958, p. 5). For generations of Labour activists, and for the wider electorate, winning a grammar-school place had provided a route from the pit or factory to a better life. If the public was confused about Labour's equivocal words, so too, was Harold Wilson. According to the education journalist Bruce Kemble, who knew him, Wilson had assumed that there would be streaming within comprehensives, but was "betrayed" by the teaching profession "who couldn't wait to get the grammar school streaming out" (Interview, *What If?* BBC Radio 4, October 23, 1993). By the 1970s, Labour had lost any sense of definition and direction. In his recently published diary, Bernard Donoughue, head of Wilson's Policy Unit, records that when, in the middle of 1975, he alerted the prime minister to the possibility that Fred Mulley, the education secretary, "might retreat from our comprehensive policy," Wilson replied that this "might not be a bad thing" (Donoughue, 2005, entry for June 11, 1975, pp. 412–13).

Reinforcing the "grammar schools for all" message, some early comprehensive schools emphasized the importance of the "academic side," adopting house systems, appointing prefects, and allocating pupils to "streams" in which they were sometimes taught by teachers wearing gowns. Although a measure of artistic license may be present in his words, John O'Farrell has remembered his Maidenhead comprehensive of the 1970s rigidly dividing the arriving cohort into two groups, with Block One being told "you're going to be doing Latin and Block Two, you look after the rabbits" (Interview, *The Schools Lottery*, Programme One, BBC Two, March 27, 2006). Elsewhere, certain comprehensives favored setting for some subjects and mixed-ability teaching for others, while some applied mixed-ability teaching to all groups. In 1976, Margaret Maden, then the head teacher of London's Islington Green School, argued that "the good mixed-ability class can extend the pupil to his or her fullest potential in a way in which a so-called homogeneous, or streamed, class rarely does" (Maden, 1976). The historic failure to resolve what kind of teaching and learning comprehensive schools stood for still resonates. Many current parents from the middle classes are attracted to using "good" local comprehensive schools, but fear that mixed-ability teaching will hold back their children from gaining coveted university places (Power et al., 2003; Swift, 2003).

The only consistent indicator for the "success" of comprehensive schooling as a national policy in the 1960s and 1970s was school numbers. This, in turn, introduced a pressure for speedy policy change, minimal consultation, blueprints unmatched by resources, and "instant," patently noncomprehensive, comprehensives. "Interim" reorganizations, which were sometimes never revisited, permitted selective and nonselective schools to coexist within the same locality, so undermining the comprehensive principle. Grammar-school preservationists, meanwhile, condemned doctrinaire planners for introducing "botched up" schemes, involving unsatisfactory mergers of geographically distant schools, for promising, but never delivering, new building sites, and of forcing teachers trained for an entirely different type of classroom experience into premature retirement. The pluralist system of educational policy making in England and Wales allowed LEAs considerable scope to devise and finesse their own reorganization schemes. From the outset, it was found that comprehensives formed around the nucleus of a grammar school were outperforming those based on secondary modern amalgamations (Tyack and Poster, 1958, p. 62), while later research using data from the 1960s and 1970s identified few "pure" comprehensives (Monks, 1968; Kerckhoff et al., 1996). Today, schools maintaining the descriptor "comprehensive" stake their claim to comprehensiveness in diverse ways. The London Oratory, the undoubtedly excellent Roman Catholic school to which Tony Blair sent his children, has been described as a comprehensive "in the same sense that the Queen is an old-age pensioner living in Westminster" (Walden, 1998).

What Were the Successes and Failures of British Comprehensive Schooling?

Like the secondary modern schools before them, the earliest comprehensive schools required a leap of faith, having been planned by politicians and officials who were far from certain to use them personally. From a twenty-first century perspective, it seems extraordinary that the debates of the 1960s and 1970s were so dominated by talk of secondary education structures, with such terms as standards, curriculum, and pedagogy barely featuring. Moreover, the central resources granted to accomplish secondary reorganization were mean and efforts to research the advantages and effectiveness of the policy were minimal.

In 1974 the education secretary Reginald Prentice associated a 28 percent rise in Advanced Level passes and an 11 percent increase in Ordinary Level passes during the period 1965–1972 with a rise in the number of English and Welsh comprehensive schools over the same period from 221, serving 6 percent of the secondary age cohort, to 1,602, serving 47 percent (*The Times*, July 4, 1974, p. 8). It is surprising, perhaps, that more champions of comprehensive schools have not come forward with similar statistics. Since the early 1960s, for example, the percentage of school leavers experiencing higher education has risen significantly (National Committee of Inquiry into Higher Education, 1997, p. 21), with increasing numbers coming from state schools, the majority of which were, and for some official purposes, still are, classified as comprehensives. Yet, while broadcasters rarely experience difficulties in finding politicians and celebrities to talk about their experiences of school selection, prominent comprehensive school alumni have been strangely silent. William Hague, the former Conservative Party leader, who made his first major speech as a schoolboy aged 16 at the 1977 party conference, has signally failed to commend the Rotherham comprehensive that prepared him for Oxford University and a career in frontline politics. In 2001, he predicted that it would not be Tony Blair, but himself, "the comprehensive school-educated leader of the Conservative Party who will end the monolithic comprehensive school system" (Hague, 2001).

The belief, once apparent at an international level (e.g., Organization for Economic Co-operation and Development, 1967) that comprehensive schools could establish common cultures to combat social inequalities, proved fanciful in Britain, though some writers have attributed advances in educational equality to comprehensive schools (e.g., Barker, 1986; Chitty, 1987; Pring and Walford, 1997). By contrast, other commentators would claim Eric James's prophesy, made 60 years ago, that comprehensive schools would precipitate "grave social, educational, and cultural evils which may well be a national disaster" (quoted in Rubinstein and Simon, 1969, p. 37), as a modern truth. Speaking in 1993, Sir Rhodes Boyson, the former comprehensive school head teacher, Black Paper editor and conservative education minister, associated comprehensives with a "slow decline in general culture" since World War II. If a balanced system of grammar schools, technical schools, and apprenticeships had been created in the postwar years, he maintained, "I don't think we would ever have moved to comprehensive schools and the lumpen proletariat we have created" (BBC Radio 4, *What If?* October 23, 1993).

Inadequately robust data has long prevented reliable comparisons of selective and nonselective education in England and Wales (Crook et al., 1999), so judgments as to the successes and failures of comprehensive schools rest upon subjective opinions. In a climate of continuing anxieties about the quality of inner-city education, these abound. In 1996, Simon Jenkins, a former *Times* editor, whose warm views about comprehensive schools contrast with most other newspaper columnists, wrote: "There are bad secondary schools, but nothing as bad as before 1965, or as bad as the 'sinks' that would result from the present opt-out policy. Heaven knows how the British workforce would look had we stayed with 11-plus selection over the past 30 years, or if we were to go back to it now" (Jenkins, 1996).

Such arguments cut no ice with Melanie Phillips, who has recently argued that to deny that comprehensive schools are responsible for school failure is "a bit like saying that a restaurant with a filthy kitchen has nothing to do with the food poisoning it gives its patrons" (Phillips, 2006).

The reputation of English comprehensive schools has undoubtedly been tarnished by several well-documented episodes. The closure of Risinghill Comprehensive, just five years after the London County Council had established it, precipitated concerns about the formation of large comprehensive schools in unsuitable buildings, as well as casting doubt upon the progressive leadership style of its head teacher, Michael Duane (Berg, 1968; Limond, 2002). Ten years after its foundation, the flagship Kidbrooke School also found itself at the heart of a controversy. Comments by a music teacher about a "difficult element" among Kidbrooke girls provided ammunition for the fiercely anti-comprehensive editor of the *Times Educational Supplement* to assert a link between unsegregated schooling and poor disciplinary standards (Kerckhoff et al., 1996, p. 67), notwithstanding the teacher's protestation that her words had been misrepresented (*The Times*, December 30, 1964, letter by Joyce Lang, p. 9). In the following decade, several television reports and documentaries, supposedly depicting "typical" classroom situations and disciplinary problems, but invariably involving footage from London secondary schools lacking a cross-section of pupil abilities, inflicted further harm upon the comprehensive movement (Chitty, 1989, p. 66).

Signifying some second, or rather third, thoughts about comprehensive schools, Harry Rée admitted in 1974 that too much had been expected of them, particularly in respect of social mixing: "We know now—and we knew then, really—that in widely different neighbourhoods there'd be widely different schools" and that "a common system doesn't produce a common result" (Institute of Education, University of London, Harry Rée papers, Box 2, 6/11, "Objectives for comprehensive education, typescript and handwritten notes for a talk at the South Shields Teachers' Center, February 1974). A recent education secretary Ruth Kelly has similarly conceded that too little thought was devoted in the 1960s and 1970s to the mission of comprehensive schools, and "what it meant to provide a high-quality education once children were inside the school gate." Interestingly, at the same time, Kelly endorsed the concept of "genuine comprehensive *education*" and confirmed that the "comprehensive *ideal*" remains powerful (*Guardian*, March 30, 2005, p. 4).

Differentiation between support for secondary comprehensive schools and for comprehensive education, which frequently embraces the primary and post-compulsory age phases, has become more evident in the past two decades. Symbolically, in 1991, *Forum for the Discussion of New Trends in Education*, the flagship journal for promoting comprehensive schools, cofounded by Brian Simon in 1958, was re-titled *Forum for Promoting 3–19 Comprehensive Education*. Five years later, its coeditor, Clyde Chitty, together with Caroline Benn, claimed for comprehensive education a number of victories, including the national curriculum, a common framework of assessment, improved progression rates to post-compulsory, and higher education and greater equality of opportunity (Benn and Chitty, 1996, pp. 461–502). Geoff Whitty subsequently pointed to the success of comprehensive education in promoting such values as inclusion, tolerance, and democracy, also highlighting academic and other achievements of schools still self-identifying as comprehensives (Whitty, 2004).

We live in strange times, however. Sir Cyril Taylor, chairman of the Specialist Schools and Academies Trust, recently told a parliamentary committee that "the specialist schools movement . . . is about comprehensive education" and that the head teachers of these schools, which rarely serve clearly defined neighborhoods and, in some cases, select a proportion of pupils, "passionately support the concept of comprehensive education" (evidence to House of Commons Select Committee on Education and Skills, December 12, 2005, question 420). The one-time view that all secondary-age children should attend their local school has now become an historical curiosity and we are left to contemplate whether comprehensive education can prosper in the twenty-first century without the presence of comprehensive schools.

Conclusion

There is a certain attraction in applying medical analogies to the case of English and Welsh comprehensive schooling. This patient began to breathe unaided after World War II, having spent two decades in an incubator. During the 1960s, the comprehensive school benefited from intensive care, but its Labour doctors administered too little oxygen and stopped short of prescribing the expensive treatment that might have enabled it to flourish. Relatively healthy in parts—its Welsh left arm (as well as a Scottish head) and English rural and small-town body parts were located away from the vital urban organs—the patient became stronger, but did not enjoy the life that was once predicted. The 1970s witnessed a division of opinion about its treatment and, in the following decade, it found itself on the books of a consultant, Margaret Thatcher, who had previously been an unsympathetic junior doctor. Yet the comprehensive school was neither surreptitiously suffocated nor subjected to a lethal injection. Instead, it was ignored, having become less interesting to those upon whom it had previously depended. Suffering from memory loss and schizophrenia, the comprehensive school was omitted from the ward rounds, became an outpatient, and went missing. From 1997, a new Labour medical team supplied other patients—foundation schools, faith schools, specialist schools, and academies—with the kind of attention and medication that comprehensive schools once craved. Though sightings are sometimes reported, the comprehensive school is missing, presumed dead. It may, however, be of some consolation that its distant cousin, comprehensive education, remains alive.

References

Barker, B. (1986). *Rescuing the comprehensive experience.* Milton Keynes: Open University Press.
Benn, C. (1992). Common education and the radical tradition. In A. Rattansi and D. Reeder (eds.), *Rethinking radical education* (pp. 142–65). London: Lawrence and Wishart.
Benn, C. and Chitty, C. (1996). *Thirty years on. Is comprehensive education alive and well or struggling to survive?* London: David Fulton.
Berg, L. (1968). *Risinghill: Death of a comprehensive school.* Harmondsworth: Penguin.
Board of Education (1938). *Secondary education with special reference to grammar schools and technical high schools* (Spens Report). London: HMSO.
——— (1943). *Educational reconstruction* (White Paper, Cmd. 6458). London: HMSO.
Butt, R. (1978a). Are the Tories ready to go comprehensive? *The Times*, March 16, p. 18.
——— (1978b). Cutting the monster schools to size. *The Times*, March 23, p. 16.
Chitty, C. (Ed.) (1987). *Redefining the comprehensive experience.* London: Institute of Education.
——— (1989). *Towards a new education system. The victory of the new right?* London: Falmer.
Costello, E. (1980). "Secondary modern," from the album *Get Happy!!* F. Beat.
Cox, C., Dyson, B., and Dyson, A. (Eds.) (1969a). *Fight for education: A black paper.* London: Critical Quarterly Society.
——— (1969b). *Black paper two.* London: Critical Quarterly Society.
Crook, D. (1993). Edward Boyle: Conservative champion of comprehensives? *History of Education* 22 (1): 49–62.
——— (2002). Local authorities and comprehensivisation in England and Wales, 1944–1974. *Oxford Review of Education* 28 (2 and 3): 247–60.
Crook, D., Power, S., and Whitty, G. (1999). *The grammar school question. A review of research on comprehensive and selective education.* London: Institute of Education.
Crosland, S. (1982). *Tony Crosland.* London: Jonathan Cape.
Davis, E. (2005). Who are you calling "bog-standard?" *Daily Telegraph*, August 31, p. 21.
Dean, D. (1998). Circular 10/65 revisited: The Labour government and the "comprehensive revolution" in 1964–1965. *Paedagogica Historica* 34 (1): 63–91.
Department for Education and Skills (2005). *Higher standards, better schools for all. More choice for parents* and pupils. (White Paper, Cm 6677). London: The Stationery Office.
Department of Education and Science (1965). *The organization of secondary education* (Circular 10/65). July 14. London: DES.
——— (1970). *The organisation of secondary education* (Circular 10/70). June 30. London: DES.
——— (1974). *The organisation of secondary education* (Circular 4/74). April 16. London: DES.

Donoughue, B. (2005). *Downing Street Diary. With Harold Wilson in No. 10*. London: Jonathan Cape.
Education Act 1944 (7 & 8 GEO. 6, c. 31), London: HMSO.
Hague, W. (2001). I, not Tony Blair, will end the comprehensive school system. *Daily Telegraph*, February 14, p. 28.
Haydn, T. (2004). The strange death of the comprehensive school in England and Wales, 1965–2002. *Research Papers in Education* 19 (4): 415–32.
Jackson, M. (1978). Conclusion by parliamentary under-secretary of state for education and science. In Department of Education and Science, *Comprehensive education. Report of a conference held at the invitation of the secretary of state for Education and Science at the University of York on 16/17 December 1977*. London: HMSO, pp. 13–14.
Jenkins, S. (1996). Tough on hypocrisy? *The Times*, January 24, p. 1.
Kerckhoff, A., Fogelman, K., Crook, D., and Reeder, D. (1996). *Going comprehensive in England and Wales. A Study of uneven change*. London: Woburn Press.
Labour Party (1958). *Learning to live. Labour's policy for education*. London: Labour Party.
Limond, D. (2002). Risinghill and the ecology of fear. *Educational Review* 54 (2): 165–72.
London County Council (1947). *London school plan. A development plan for primary and secondary education*. London: LCC.
Lovett, T. (1958). Educational opportunities. (C) More advanced courses and the sixth form. In National Union of Teachers, *Inside the comprehensive school. A symposium contributed by heads of comprehensive schools in England and Wales* (pp. 48–60). London: The Schoolmaster Publishing Company.
McCulloch, G. (2000). Publicizing the educational past. In D. Crook and R. Aldrich (eds.), *History of education for the twenty-first century* (pp. 1–16). London: Institute of Education.
Maden, M. (1976). Why there should be more for everyone in comprehensives. *The Times*, January 28, p. 11.
Ministry of Education (1947). *Organisation of secondary education. Further considerations suggested by development plan proposals* (Circular 144). London: HMSO.
Monks, T.G. (1968). *Comprehensive education in England and Wales. A survey of schools and their organisation*. Slough: National Foundation for Educational Research.
National Committee of Inquiry into Higher Education (1997). *Higher education in the learning society* (Dearing Report). London: HMSO.
Norton, P. (1985). Introduction. In Philip Norton (ed.), *Parliament in the 1980s* (pp. 1–19). Oxford: Basil Blackwell.
Organization for Economic Co-operation and Development (1967). *Social objectives in educational planning*. Paris: OECD.
Phillips, M. (2006). Selection is the *only* way to save our schools. *Daily Mail*, January 23, p. 14.
Power, S., Edwards, T. Whitty, G., and Wigfall, V. (2003). *Education and the middle class*. Buckingham: Open University Press.
Pring, R. and Walford, G. (Eds.) (1997). *Affirming the Comprehensive ideal*. London: Falmer.

Rée, H. (1956). *The essential grammar school*. London: Harrap.
Rubinstein, D. and Simon, B. (1969). *The evolution of the comprehensive school, 1926–1972*. London: Routledge & Kegan Paul.
Simon, B. (1991). *Education and the social order, 1940–1990*. London: Lawrence & Wishart.
St. John-Stevas, N. and Brittan, L. (1975). *How to save your schools*. London: Conservative Political Centre.
Swift, A. (2003). *How not to be a hypocrite. School choice for the morally perplexed parent*. London: Routledge.
Tyack, N.C.P. and Poster, C.D. (1958). External examinations and internal assessments. In National Union of Teachers, *Inside the comprehensive school. A symposium contributed by heads of comprehensive schools in England and Wales* (pp. 61–72). London: The Schoolmaster Publishing Company.
Vernon, P.E. (Ed.) (1957). *Secondary school selection*. London: Methuen.
Walden, G. (1998). The big lie about scrapping grammars. *Evening Standard*, November 17, p. 11.
Whitty, G. (2004). Developing comprehensive education in a new climate. In M. Benn and C. Chitty (eds.), *A tribute to Caroline Benn: Education and democracy* (pp. 97–110). London: Continuum.
Wilson, A.N. (1996). So, is it all Shirley Williams fault? *Evening Standard*, June 25.
Yates, A. and Pidgeon, D.A. (1957). *Admission to grammar schools*. London: Newnes.

Chapter 10

The Comprehensive Ideal in New Zealand: Challenges and Prospects

Gregory Lee, Howard Lee, and Roger Openshaw

In post–World War II New Zealand, comprehensive schooling gradually became the dominant model of universal secondary education. As in other countries, this preferred model has faced several major challenges wherein its ideals and outcomes have undergone considerable critical scrutiny. This chapter begins by looking at the genesis of the comprehensive ideal during the interwar period as it arose to head off the rival notion of selective secondary schools. The chapter then traces its development during the early postwar years when comprehensive schooling experienced several contradictory forces, including the comprehensive ideals of the 1942 *Consultative Committee on the Post-Primary School Curriculum* (Thomas Committee), continuing debates over academic standards, and the economic pressures of the baby boom era. The chapter goes on to illustrate how, from the early 1970s, comprehensive schooling faced contradictory political and philosophical imperatives. These included calls for more curriculum relevance, demands for inclusion and equity, mounting fiscal pressures, and growing pressure for competition and choice, culminating in the (Picot) *Report of the Taskforce to Review Education Administration* (Department of Education, 1988a), *Tomorrow's Schools* (Department of Education, 1988b), and a radically reshaped education system. The chapter concludes that, especially since 1990, debate over the comprehensive ideal has not only been recast but also has intensified.

The Dawning of the Comprehensive Ideal

The Hogben-Seddon free place system, which offered free secondary education to all primary school leavers who had passed the Standard 6 (Form 2/Year 8) Proficiency examination from 1901, was designed to furnish secondary education for deserving students (Lee, 1991; McKenzie, Lee, and Lee, 1996; Openshaw, Lee, and Lee, 1993). The system's popularity, however, rapidly outgrew the conservative projections of its author, Inspector-General of Schools, George Hogben (Lee, 2005; Murdoch, 1944). Hogben's educational philosophy was not founded on a fully fledged mass schooling model for comprehensive postprimary schooling. Instead, from the outset, he favored having some discernible differentiation between academic, town-based, secondary schools and the higher (secondary) departments of the rurally situated district high schools (Lee, 2005; Thom, 1950).

It was this dual function—to offer both academic and "vocational" (i.e., "nonacademic") instruction—that persuaded Hogben to accept that district high schools were operating as comprehensive institutions, even if not in name (Lee, 2005; Thom, 1950). He had long been convinced that *both* secondary and district high schools could and should give their students a general education that did not have to be defined and dominated by an academic, examination-oriented curriculum. It was this philosophy that encouraged him to advocate a multilateral postprimary schooling model (i.e., where academic and nonacademic courses would both be offered within the one institution) toward the end of his tenure as Inspector-General of Schools (Openshaw, Lee, and Lee, 1993). The resultant common core curriculum would promote general education and allow optional studies to be chosen "to meet students' own needs" (Openshaw, Lee, and Lee, 1993, p. 107).

Hogben's ideas were not accepted universally. Josiah Hanan, minister of education (1915–1919), believed that the solution lay with more institutionally based curricular differentiation (Murdoch, 1944; Openshaw, Lee, and Lee, 1993). Hanan's advocacy of institutional differentiation, in conjunction with a small compulsory common core curriculum, was arguably the most important feature of his term of office (Openshaw, Lee, and Lee, 1993). It was also to dominate Christopher J. Parr's tenure as minister of education (1920–1926) more overtly (Openshaw, 1995). Parr was determined to institute a selective schooling policy beyond the primary sector (McKenzie, 1987). He was, however, forced to concede during his term as minister of education that the postprimary schools were in fact expanding their curricular offerings, in an effort to meet the perceived and real "needs" or requirements of their students (Openshaw, Lee, and Lee, 1993) and to acknowledge that in this regard school authorities were often exercising their own initiative, independent of educational legislation.

By the late 1920s an uneasy and obvious political tension remained over the advocacy of subject- and gender-based differentiation and the prescription of a limited set of "general education" subjects. In the absence of clear government direction, educationists continued to explore alternative structures for postprimary schooling at the same time as the New Zealand Labour Party was publicly opposing any adoption of a selective postprimary schooling policy. The party's education spokesperson, Peter Fraser—a working-class Scotsman whose own primary schooling had been terminated early by the need to contribute to his family's income—argued that to base access to schooling on the country's existing social class structure meant that a sizeable group of students (i.e., working-class youth) would receive only narrow technical and other directly vocational training whereas others (i.e., children of the middle class) would benefit from "a full cultural education" (*NZPD*, 1923, p. 462; Openshaw, Lee, and Lee, 1993, pp.153–54).

Fraser's curriculum philosophy—indeed, that of the Labour Party—was warmly endorsed by Frank Milner, the prominent but occasionally controversial Rector of Waitaki Boys' High School (Oamaru) and frequent commentator on a host of education and social issues (Lee and Lee, 2002; Openshaw, Lee, and Lee, 1993). Milner had advocated a secondary school curriculum philosophy that involved "a harmonious combination of the cultural and the practical and economic in one organic whole" (Openshaw, Lee, and Lee, 1993, p. 155; Tate, 1925, p. 120); essentially a compulsory general education curriculum to be followed by some specialized studies based on optional subjects chosen by students, with parental and teacher input. The Fraser et al. nonselective, non-differentiated postprimary schooling policy slowly gained momentum to the extent that by the time the Labour Party was first elected to government in December 1935, the New Zealand public was well aware of the party's manifesto on education and other matters (Paul, 1946). The New Education Fellowship (NEF) Conference held in New Zealand in 1937 was undoubtedly influential in this regard. Peter Fraser, now minister of education, reported, for example, that this Conference "marked the commencement of an educational renaissance from which much will come" (Fraser, 1938, p. ix).

Two years after the NEF Conference, Fraser issued the Labour Government's most detailed policy statement on education. In this document—drafted in conjunction with the newly appointed assistant director of education, Dr C.E. Beeby—Fraser lambasted the long tradition of institutionally selective schooling in the New Zealand postprimary sector and emphasized the electorally appealing, elastic notion of equality of educational opportunity for all youth (*AJHR*, 1939; D. McKenzie, personal communication, September 17, 2005). While there was no direct indication of the government's intentions with particular reference to either comprehensive schooling or an expanded general education curriculum,

Fraser did reveal a desire to provide students with "post-primary education of a kind for which he [or she] is best fitted" (*AJHR*, 1939, p. 3). Also significant, however, was his statement that equal educational opportunity did not guarantee that students "shall inevitably have exactly the same education in every detail" (p. 6). Consequently curriculum "adaptation" was destined to figure prominently in the government's education policy, in light of Fraser's assumption that the school leaving age would be raised to 15 years in the near future and in response to the abolition of the Standard 6 (Form 2/ Year 8) Proficiency examination in 1937 (Openshaw, Lee, and Lee, 1993).

Fraser's tenure as minister of education (1935–1940) coincided with Beeby's appointment as assistant director (1938–1939) and then director of education (1940–1960). Beeby envisaged that the introduction of any curriculum reform would prove difficult, because of the conservative legacy of historical attitudes, traditions, and competition associated with the establishment of different types of high schools (Beeby, 1937). His pessimism was to prove well founded (see, e.g., Alcorn, 1999, p. 123). Irrespective of what might transpire, however, Beeby sought to avoid a "mechanical copy of overseas models" in the belief that New Zealand should find "[its] own solution of her own problems" (Beeby, 1939, p. 702).

Beeby was not the only educationist in this period who referred to the "delicate operation" (Beeby, 1937, p. 238) associated with institutional change and the need to give very careful consideration to postprimary curriculum matters. In the mid-1930s Frank Milner had proposed a compulsory common core curriculum for *secondary* school students (Lee and Lee, 2002; Openshaw, Lee, and Lee, 1993)—one he deemed "[suitable] to our national needs . . . and our distinctive conditions of life" (Milner, 1936a, p. 11)—that embraced a range of aesthetic, science, social science, and humanities subjects well beyond those specified in the current free place legislation. These curriculum ideas won universal acclaim from participants at the New Zealand Secondary Schools' Association (NZSSA) annual conference in May 1936 (Milner, 1936b; see also Campbell, 1941).

What is seldom emphasized in historical accounts of New Zealand postprimary schooling in the 1930s and 1940s is the fact that Milner intended his curriculum model to apply to *secondary schools* only (Milner, 1936a). Thus, when he referred to the "secondary school curriculum" Milner meant precisely that. When comparisons are made, as they ought to be, between Milner's proposal and the Thomas Committee's recommendations, it is evident that the latter expressly intended to extend their suggested compulsory common core curriculum to *all* types of postprimary institutions. This entailed going well beyond Milner's idea of translating secondary schools into comprehensive institutions: under the "Thomas" model, technical high and secondary departments of district high schools—as well as registered (private) secondary schools (Department of Education, 1959)—were obliged to undergo this transformation. Recognizing such a distinction, we believe, is essential to gaining an appreciation of subsequent (post-1946) moves to further consolidate comprehensive schooling philosophy and practice within the postprimary sector. This signaled a definite departure from the British Spens and Norwood Reports (Board of Education, 1939, 1943; see also Mason, 1944), with their advocacy of rigid institutional differentiation (Lee, 1991).

The Thomas Committee's report (1944) has deservedly been the subject of considerable discussion by historians of New Zealand education and other educationists. Some have noted the Committee's advocacy of Milner's common core secondary school curriculum, and have pointed to many difficulties associated with the Committee's attempts to formulate a revised School Certificate examination and to reconcile this with a modified compulsory general education curriculum (Alcorn, 1999; Department of Education, 1959; Lee and Lee, 1992, 2002; Lee, O'Neill, and McKenzie, 2004; McKenzie, 1983; Openshaw, Lee, and Lee, 1993). The Committee was insistent that regardless of the type of postprimary school pupils attended—and irrespective of their academic and other abilities and occupational aspirations—all students would be catered for through a (variable) common core curriculum and through new courses in schools (Department of Education, 1959). The "core studies" specified by the Committee were English language and literature, social studies, elementary mathematics, general science, music, one craft or a fine art, and physical education, although no courses were prescribed. Recognition of and provision for individual differences loomed large in the Committee's thinking, which culminated in the recommendations that while the general education curriculum had to represent "a generous and well balanced education" and to prepare an adolescent "for an active place in our New Zealand society as worker, neighbour, homemaker, and citizen" (p. 5), the postprimary institutions nevertheless should be "[free] to develop courses in terms of their own requirements" (p. 13).

The Labour Government's second minister of education, Rex Mason, reacted positively to the Thomas Report when he received it in November 1943 (see, e.g., Mason, 1944). Yet Mason still sought to clarify policy about the future direction of postprimary schooling, and a wide variety of educational issues, with the publication of *Education Today and Tomorrow*, the Labour Government's blueprint for education reform, in late 1944, intended initially for distribution at a national education conference that year. He rightly concluded that raising the school-leaving age to 15 years in 1944 signaled far more than "a mere quantitative expansion" (p. 39) in the postprimary sphere, for it marked a significant change in the *very nature* of the schooling to be provided. Accordingly, a five-year settling-in period was specified to allow teachers and the Department of Education staff to adjust to the new curriculum regulations and requirements in the wake of the Thomas Report and to "[make] the new system work satisfactorily" (p. 45). Both Beeby and Mason, however, had been overly optimistic in predicting that postprimary teachers in particular would be fully conversant with, and sympathetic to, the comprehensive schooling philosophy and the specific curriculum suggestions outlined in the Thomas Report.

Challenges to Comprehensive Schooling: From Policy to Practice

Reflecting on the Thomas Report and its legacy some 45 years later, Beeby (1992) acknowledged that the requirement for all postprimary schools post–World War II to operate as multilateral institutions had been an especially onerous one for staff and students. Consequently, the process of reforming the entire schooling system had taken "a matter not of years but of decades and even generations" (p. 179; see also Thom, 1950). Beeby's preferred policy—and that of successive Labour government ministers of education (e.g., Mason and Terence McCombs)—was for the comprehensive schooling philosophy to be adopted progressively by individual postprimary institutions as their staff began designing new courses and determining the subject matter of the new common core curriculum (*AJHR*, 1948). The subsequent curricular liberalization would assist the translation of these schools into comprehensive organizations, he envisaged (see, e.g., Beeby, 1992). It also meant that the public would come to appreciate that the government's equality of educational opportunity objective did (and would) not guarantee identical treatment of pupils. Some kind of pupil selection was thus destined to remain a feature of postprimary schooling, despite Beeby's declaration that New Zealand provided "an example of secondary education without selection" (Beeby, 1956, p. 396). He was, nevertheless, adamant that the New Zealand high school reforms were markedly different from those of contemporary Britain, and declared that "New Zealanders would not willingly accept the re-imposition of a system of selection for secondary schools and the consequent [English] tripartite school system" (Beeby, 1956, p. 399). Numerous education reports, articles, and documents released over the next 30 years echoed Beeby's sentiments and signaled New Zealanders' clear preference for a more inclusive, less selective secondary school curriculum.

In a 1973 letter to the author of a soon-to-be-published article on the Thomas Report (Whitehead, 1974), the then director of education, W.L. Renwick, observed that few people were in a position to appreciate either the extent of the controversy that had surrounded the implementation of the Thomas Report or the degree to which allegations of declining academic standards—ostensibly due to the introduction of a mass comprehensive education system—came to be laid squarely at the door of the early postwar Department of Education (Openshaw, 2003). As early as 1944, employers were criticizing academic standards and teaching methods publicly (Lee and Lee, 1992; Openshaw, Lee, and Lee, 1993). Moreover many university academics lamented what they took to be the passing of "traditional" prewar grammar school education (Openshaw, 2003), while Catholic educators generally remained profoundly suspicious of the secondary education reforms for decades to come (Collins, 2005). Public concern and media interest in the question of academic standards was, therefore, comparatively high in the early postwar years.

Reacting to such criticism, the 1962 Currie Commission—an 885-page 16-person Commission appointed by the minister of education, Phil Skoglund, to inquire into the overall efficiency and effectiveness of the New Zealand state education system—came out firmly in favor of comprehensive schooling because "[it] meets the wishes of the people of New Zealand" (p. 217). It did, however, concede that the model was under attack, on the grounds that it was seen publicly as both catering for academic mediocrity and encouraging uniformity (Commission on Education in New Zealand, 1962). Whilst they praised comprehensive schools for allowing students to move to a variety of occupations "without distinction" the Commission also observed that multilateral schools could still select, group, and thus differentiate between students *internally*, usually by allocating them to different courses (p. 220).

The Currie Report has been regarded as an attempt to reiterate the acceptability of the Fraser-Beeby common core curriculum (comprehensive schooling) model (G. Lee, 2003). Notwithstanding this effort, however, during the period from the mid-1970s to the end of the 1980s comprehensive schooling faced further dilemmas as New Zealand entered an era characterized by both rapid social change and increasing economic uncertainty. Whilst the popular press and employers, supported by some politicians, continued to criticize low academic standards liberal secondary school teachers began to question the continuing public emphasis on the academic curriculum, with its domination by external examinations and its relentless subject-discipline rigidity. They urged the New Zealand Post-Primary Teachers' Association (NZPPTA) to initiate the first thorough examination of secondary school system objectives since the publication of the Thomas Report (New Zealand Post-Primary Teachers' Association, 1969). The resulting publication, *Education in Change*, produced by the NZPPTA focused on the need for education to be viewed as a process of individual and social growth rather than as a narrow form of training people for material rewards. It also emphasized the importance of responding to economic and social changes such as the desirability of supplementing the export of primary produce by high-quality manufacturers produced by a highly qualified, knowledgeable, and flexible workforce, and noted changes to family structures, women, the growth of science, technology, and the mass media, together with educational changes—for example, the huge increase in sixth-form (Year 12) enrolments and modifications to the type of secondary schools where coeducational institutions were the norm rather than the exception. The report concluded that schools now offered a wider range of courses than did the predominantly academic schools of the 1940s (NZPPTA, 1969).

The Curriculum Review documents of 1984 and 1986 revealed this polarization dramatically. Set up by the National Government's deeply conservative minister of education, Merv Wellington, the *Core Curriculum Review* (1984) identified curriculum overlap and overload as major problems for secondary schools (Department of Education, 1984). Pointing to the pressing need for a more highly skilled workforce, the Review recommended an increase in the minimum time allocation for their suggested new core curriculum to 70 percent of the total annual time available, with the minimum time for English, social studies, mathematics and science instruction in Form 3 raised from the present 627 hours to 1,000 hours annually, in direct recognition of the importance of these fundamental studies. The increased time for mathematics and science was justified in terms of the growing importance of technology in the modern world. Following a change of Government, Labour's liberal minister of education, Russell Marshall, established another curriculum review committee, which recommended that the old core curriculum be revoked in favor of a new national curriculum common to all schools from new entrants (Year 1) to the end of form 5 (Year 11). Form 6 and 7 (Year 12 and 13) courses were to be developed as a "continuation and extension of the common core curriculum" (Department of Education, 1986, p. 128). The new curriculum was to represent a "broad and general education," and was to consist of specific curriculum principles and aspects of learning (knowledge, skills, attitudes, and values). Each school was to have a curriculum planning group responsible to the school's managing body for developing a school scheme consistent with the national common curriculum. Thus schools were to be held accountable nationally, in terms of the national principles, and locally, in accordance with the needs of the school's community.

Ongoing curriculum controversies, however, were to some extent overshadowed by a growing disenchantment with education itself, including the claim that the comprehensive ideal was the institutional expression of "equality of opportunity". From the mid-1970s academic radicals, influenced by "the new sociology," began to mount a powerful critique of the education system for allegedly disadvantaging Maori, other minorities, and women. By the early 1980s, this criticism was reflected increasingly in publications whose authors claimed that the failure of postwar educational reforms undercut the liberal myth of progress (Openshaw, 1995). The old consensus had collapsed and the opportunity had come to establish a new (radical) alternative (Codd, Harker, and Nash, 1985; Openshaw, 1995). Under this new system, Maori parents and the parents of other disadvantaged groups would be able to "shop around" for better options (Walker, 1985). Nevertheless schools remained the site of a pervasive hegemony, designed to domesticate teachers through regulation, disciplinary procedures, and a prescribed curriculum (Ramsay, 1985). Many of these writers were to have a major influence during the 1980s and beyond, with Walker continuing to exercise a national role in Maori agitation for the recognition of initiatives such as Kohanga Reo (Maori preschools where only the Maori language is spoken) and Kura Kaupapa (Maori language immersion primary schools), and with Ramsay being appointed to the Picot Committee.

Meanwhile the mounting fiscal crisis, coupled with rising youth unemployment, prompted growing political demands for educational reform on both economic and social efficiency grounds (H. Lee, 2003). A series of adverse Treasury reports saw the Department of Education, a key ally of comprehensive education, publicly criticized in Cabinet and in the national press for wasteful expenditure (Openshaw, 2003). Backed by Treasury, Derek Quigley, associate minister of finance in R.D. Muldoon's National Government, prepared a confidential report to Cabinet in 1980 on public expenditure in which he recommended that all government departments, including Education, examine their expenditures in the light of government policy objectives and existing resources rather than in terms of their supposed desirability. Educational expenditure continued to be singled out as being both wasteful and inefficient, resulting in further adverse media comment and consequent reductions in public confidence (Openshaw, 2003).

Released prior to the Picot Report, the Probine-Fargher Report (March 31, 1987) on the management, funding, and organization of continuing education in the polytechnic sector concluded that New Zealand polytechnics were excessively dominated by the central Department of Education, had low status, and were poorly resourced. Reflecting the new ideological mix of left-wing disenchantment and neoliberal criticism with the education system, Probine and Fargher adopted a broad view of continuing education and vocational training in the foreword to their report when they outlined what they perceived as being "significant issues":

> [The] low participation rates in tertiary education in New Zealand compared with other countries; the rate at which people with technological skills are being produced; a highly centralised style of management which inhibits the adoption of an entrepreneurial approach to the delivery of services; lack of coordination in the delivery of training; and last but by no means least, the need to provide special help for socially disadvantaged groups such as young unemployed, Maori who have been disadvantaged by a pakeha-based (i.e., white, non indigenous) education system, women who wish to enter non-traditional occupations, residents of rural areas, the disabled, and the socially isolated. (Probine and Fargher, 1987, Foreword)

The Picot Report (1988) and the subsequent policy document, *Tomorrow's Schools* (1988), that radically restructured the education system at all levels from primary schools through to universities was not concerned with comprehensive secondary schooling as such. However, the comprehensive ideal of equality through diversity was challenged first, by much of the rhetoric that lay behind the reforms and, second, by the educational environment that had emerged by the end of the 1980s as a result of profound social, economic, and political changes. The Picot Committee began with a "blank slate," albeit one in which the existing system of educational administration was to disappear in its entirety rather than undergoing substantial reform. The result was to be the transformation of schools into autonomous, self-managing units, competing for students in an economic marketplace characterized by consumer choice. Existing bureaucratic structures were replaced by centralized auditing, regulatory and monitoring processes across the entire education sector (H. Lee, 2003; Lee, O'Neill, and McKenzie, 2004).

From 1990 under a newly elected National Government, the new educational policy environment witnessed a raft of curriculum changes in response to the eight-level *New Zealand Curriculum Framework* (NZCF) (Ministry of Education, 1993) that provided a structural template of seven "essential learning areas" (language and languages, mathematics, science, technology, social studies, the arts, and health and physical well-being) and eight "essential skills" (communication skills, numeracy skills, information skills, problem-solving skills, self-management and competitive skills, social and cooperative skills, physical skills, and work and study skills). This framework applied to all students in primary (Years 1–8) and postprimary (Years 9–13) schools. A major outcome was the effective reversal of the Thomas Report's emphasis on a "generous and well-balanced education" that emphasized personal and social as well as economic goals in favor of a new emphasis on a training culture that stressed economic and competitive imperatives and outputs (Lee, O'Neill, and McKenzie, 2004). In the first years of the new millennium the Labour Government under Prime Minister Helen Clark has continued to endorse the contradictory aims of the NZCF, with little acknowledgment that goals such as competition and cooperation, knowledge and skills, are in any way antithetical (Lee, Hill, and Lee, 2004). Official documents and pronouncements thus actively promote aspects of a utilitarian educational philosophy whilst at the same time stressing the intrinsic benefits of education in the Beeby mold (Lee, Hill, and Lee, 2004).

Given these contradictions and the reluctance to acknowledge them, it is perhaps not surprising that there has been a renewed assault on the comprehensive ideal. The New Zealand comprehensive schooling philosophy, it is presently alleged, has produced a situation in which "as many as one in five pupils in the system is failing" and where "there is a large group at the bottom who are not succeeding" (Thomson, 2005, p. A1). In this context, Chris Saunders and Mike Williams, principals of Onehunga High School and Aorere College in Auckland, respectively, have commented, however, that having underachieving students in secondary schools in particular is *not* a recent phenomenon. A large "tail" of poor-performing high school students has long been a cause of concern, Williams contends. Notwithstanding this observation a former president of the Mangere (Primary) Principals' Association, Keith Gayford, has promptly laid responsibility for this problem squarely on the secondary school sector. Claiming that the high schools' outdated curriculum was primarily to blame, Gayford then confidently asserted: "Many of their programmes seem to be based on the needs of kids 20 years ago. I think you'll find it is the performance of [secondary] schools, not students, that is the problem" (p. A1).

A similar opinion has been expressed recently by the editor of *The New Zealand Herald* who declared that "the state system is designed to *produce* [italics added; *sic*] a broadly similar range of ability in all schools" ("Egalitarian approach," 2005, p. A12). Questioning "whether it is wise to ask all schools to cater to the whole range [of students]" (p. A12), this journalist recommended that specialized schools be created deliberately to provide for different types of students. By implementing a policy of institutional differentiation the disconcertingly high (New Zealand) school failure rate would be reduced considerably and "equitable results" would somehow be secured, it was presumed. Anything less, the editor concluded, was patently "an indictment of the egalitarian set-up" (p. A12).

Conclusion: Revisiting the Past? The Comprehensive Debate Renewed

As these recent press commentaries amply demonstrate comprehensive schools, like all other learning institutions, remain profoundly affected by the Picot reforms, with the ramifications continuing to be played out. Beyond this, however, the shadow of the interwar debates in New Zealand over selective versus comprehensive schooling still exercise an influence, as does the English historical experience with the Hadow (1926), Spens (1939), and Norwood (1943) committees reports (Board of Education, 1926, 1939, 1943; Kazamias and Massialas, 1965; Raynor, 1972; Simon, 1969).

Thus, whilst the ongoing New Zealand experience with comprehensive schooling is undoubtedly instructive it is not unique. In the United States, the debate over comprehensive high schools has been renewed significantly in recent years. Positioned symbolically if not ominously at the very close of the twentieth century a book by David Angus and Jeffrey Mirel, provocatively entitled *The Failed Promise of the American High School*, argued that at the heart of the American conception of good high school education are the near-sacrosanct tenets of educational progressivism—that is, that the curriculum should be differentiated according to individual needs, aspirations, and interests—and that such differentiation should be accommodated within a single institution. This remarkably resistant legacy, they claimed, has resulted in an ongoing mediocrity in academic attainment, with consequent disadvantages for working-class and minority students. Educational professionals, through their capture of the processes of curriculum making and their continuing adherence to progressivist philosophies, were depicted as constituting a serious impediment to current political calls to monitor American educational outcomes more closely, to encourage higher academic standards, and to make the system more internationally competitive (Angus and Mirel, 1999). A number of subsequent researchers (Kantor and Lowe, 2004) have, nevertheless, questioned the assumption that schools today are worse academically than they were in the past, arguing instead that rather than attempting to recapture a largely mythical past we need to confront the real inequities that have been embedded in school systems historically, and to understand how those groups long marginalized by the system have confronted those inequities in ways that have promoted greater access and improved educational quality.

The experience of the United States allows us to view contemporary critiques of New Zealand comprehensive schooling, especially in the popular press, in both their historical and international contexts. For instance, it is rarely conceded by critics that any mass schooling model would inevitably highlight considerable differences between the large numbers of students involved. Hence, the core questions that need addressing are as follows: How have these variations been catered for within schools, once they have been identified, and in precisely what institutional context? As we have shown, this debate is best understood in relation, first, to the consequences of political efforts to expand the schooling entitlement of students who left primary school with a Standard 6 examination pass, and second, to the numerous issues associated with raising the school-leaving age to 15 years in 1944 (Lee and Lee, 1992; H. Lee, 2003; Openshaw, Lee, and Lee, 1993). Few consequences of these two measures were anticipated by their respective authors.

Ian McLaren has noted that as late as the 1970s there was no clear consensus about whether teachers and school authorities had "adapted fully to post-1945 circumstances" (McLaren, 1974, p. 120). Revisiting the 1940s' postprimary reforms some 50 years later, Openshaw (1995) coined the term "unresolved struggle" to describe the legacy of the policy "settlement" of that period. Vocational and general education conflicts remain a feature of the postprimary schooling landscape in and beyond New Zealand, given the reality that teachers and school authorities are (and have been) expected to satisfy a wide variety of demands placed upon them from several quarters: economic, social, and political. As a result, debate over "exactly what place people are to be assigned to by whom, on what grounds and for what purpose" (Openshaw, 1995, p. 137) did not diminish with the introduction of comprehensive postprimary schooling, contrary to what its proponents had expected. As this chapter has demonstrated, New Zealand still struggles to comprehend the full impact of mass postprimary schooling some 60 years after its inception. This dilemma undoubtedly remains "a problem of secondary education unsolved worldwide" (Beeby, 1984, p. 89).

References

Alcorn, N. (1999). *To the fullest extent of his powers: C.E. Beeby's life in education.* Wellington: Victoria University Press.

Angus, D.L., and Mirel, J.E. (1999). *The failed promise of the American high school, 1890–1995.* New York: Teachers College Press.

Appendices to the Journals of the House of Representatives [*AJHR*]. (1927). (1939). (1948). Report of the Minister of Education. E-1.

———. (1956). Annual report of the Director of Education. E-1.

Beeby, C.E. (1937). The education of the adolescent in New Zealand. In H.V. Usill (ed.), *The year book of education* (pp. 219–38). London: Evans Brothers/University of London Institute of Education.

———. (1939). Technical education in New Zealand. In H.V. Usill (ed.), *The year book of education* (pp. 690–702). London: Evans Brothers/University of London Institute of Education.

———. (1956). New Zealand—An example of secondary education without selection. *International Review of Education* 2 (4): 396–407.

———. (1984). A problem of secondary education unsolved worldwide. In G. McDonald and A. Campbell (eds.), *Looking forward: Essays on the future of education in New Zealand* (pp. 89–111). Wellington: Te Aro Press.

———. (1992). *The biography of an idea: Beeby on education.* Wellington: New Zealand Council for Educational Research.

Board of Education. (1926). *Report of the consultative committee on the education of the adolescent* (The Hadow Report). London: His Majesty's Stationery Office.

———. (1939). *Report of the consultative committee on secondary education with special reference to grammar schools and technical high schools* (The Spens report). London: His Majesty's Stationery Office.

———. (1943). *Report of the committee of the secondary school examinations council on the curriculum and examinations in secondary schools* (The Norwood report). London: His Majesty's Stationery Office.

Campbell, A.E. (1941). *Educating New Zealand* (New Zealand centennial surveys No. VIII). Wellington: Department of Internal Affairs.

Commission on Education in New Zealand. (1962). *Report of the commission on education in New Zealand* (The Currie report). Wellington: Government Printer.

Codd, J., Harker, R., and Nash R. (Eds.) (1985). *Political issues in New Zealand education.* Palmerston North: Dunmore Press.

Collins, J. (2005). *"For the common good." The Catholic educational mission in transition, 1943–1965.* Unpublished doctoral dissertation, Massey University, Palmerston North.

Department of Education. (1959). *The post-primary curriculum: Report of the committee appointed by the minister of education in November, 1942* (Reprint) (The Thomas report). Wellington: Government Printer.

———. (March 1984). *A review of the core curriculum for schools.* Wellington: Author.

———. (July 1986). *The curriculum review.* Wellington: Author.

Department of Education. (1988a). *Administering for excellence: Effective administration in education* (The Picot report). Wellington: Government Printer.
———. (1988b). *Tomorrow's schools: The reform of educational administration in New Zealand*. Wellington: Government Printer.
Egalitarian approach fails pupils. [Editorial]. (October 27, 2005). *The New Zealand Herald*, p. A12.
Fraser, P. (1938). Foreword. In A.E. Campbell (ed.), *Modern trends in education: The proceedings of the New Education Fellowship conference held in New Zealand in July 1937* (pp. ix–x). Auckland: Whitcombe & Tombs.
Kantor, H. and Lowe, R. (2004). Reflections on history and quality education. *Educational Researcher* 33 (5): 6–10.
Kazamias, A.M. and Massialas, B.G. (1965). *Tradition and change in education: A comparative study*. Englewood Cliffs, NJ: Prentice-Hall.
Lee, G. (1991). *From rhetoric to reality: A history of the development of the common core curriculum in New Zealand post-primary schools, 1900–1945*. Unpublished doctoral dissertation, The University of Otago, Dunedin.
———. (2003). Thinking comprehensively: Some comparisons between the New Zealand Thomas report (1944) and the New South Wales Wyndham report (1958). *Education Research and Perspectives* 30 (2): 26–59.
Lee, G. and Lee, H. (1992). *Examinations and the New Zealand school curriculum: Past and present* (Delta Research Monograph No.12). Palmerston North: Massey University, Faculty of Education.
Lee, G., Hill, D., and Lee, H. (2004). The New Zealand curriculum framework: Something old, something new, something borrowed, something blue. In A.-M. O'Neill, J. Clark, and R. Openshaw (eds.), *Reshaping culture, knowledge and learning: Policy and content in the New Zealand Curriculum Framework* (pp. 71–89). Palmerston North: Dunmore Press.
Lee, G. and Lee, H. (December 2002). *Making Milner matter: Some comparisons between the Milner (1933–1936), Thomas (1944), and subsequent New Zealand secondary school curriculum reports and developments*. Paper presented at the annual meeting of the New Zealand Association for Research in Education, Palmerston North.
Lee, H. (2003). Outcomes-based education and the cult of educational efficiency: Using curriculum and assessment reforms to drive educational policy and practice. *Education Research and Perspectives* 30 (2): 60–107.
———. (2005). The New Zealand district high school: A case study of the conservative politics of rural education. *Education Research and Perspectives* 32 (1): 12–51.
Lee, H., O'Neill, A-M., and McKenzie, D. (2004). "To market, to market . . . " The mirage of certainty: An outcomes-based curriculum. In A.-M. O'Neill, J. Clark, and R. Openshaw (eds.), *Reshaping culture, knowledge and learning: Policy and content in the New Zealand curriculum framework* (pp. 47–70). Palmerston North: Dunmore Press.
Mason, H.G.R. (1944). *Education today and tomorrow*. Wellington: Government Printer.
McKenzie, D. (1983). Politics and school curricula. In W.J.D. Minogue (ed.), *Adventures in curriculum* (pp. 20–34). Sydney: George Allen & Unwin.

———. (August 1987). *The New Zealand Labour party and technical education: 1919–1930.* Paper presented at the annual meeting of the Australian and New Zealand History of Education Society, Hobart.

McKenzie, D., Lee, H., and Lee, G. (1996). *Scholars or dollars? Selected historical case studies of opportunity costs in New Zealand education.* Palmerston North: Dunmore Press.

McLaren, I.A. (1974). *Education in a small democracy: New Zealand.* London: Routledge & Kegan Paul.

Milner, F. (1936a). Secondary schools' curriculum with special reference to its suitability for our national needs. *STA: The Official Organ of the New Zealand Secondary Schools' Association and the New Zealand Technical School Teachers' Association* 3: 11–14.

———. (1936b). Report on the suitability of the secondary curriculum for modern needs. *STA*, 3, 8.

Ministry of Education. (1993). *The New Zealand Curriculum Framework.* Wellington: Learning Media.

Murdoch, J.H. (1944). *The high schools of New Zealand: A critical survey.* Wellington: New Zealand Council for Educational Research.

New Zealand Parliamentary Debates [*NZPD*]. (1923), Vol. 200. Wellington: Government Printer.

New Zealand Post-Primary Teachers' Association. (1969). *Education in change: Report of the curriculum review group* (The Munro Report). Auckland: Longman Paul.

Openshaw, R. (1995). *Unresolved struggle: Consensus and conflict in state post-primary education.* Palmerston North: Dunmore Press.

———. (2003). Preparing for Picot: Revisiting the "neoliberal" educational reforms. *New Zealand Journal of Educational Studies* 38 (2): 135–50.

Openshaw, R., Lee, G., and Lee, H. (1993). *Challenging the myths: Rethinking New Zealand's educational history.* Palmerston North: Dunmore Press.

Paul, J.T. (1946). *Humanism in politics: New Zealand Labour party retrospect.* Wellington: New Zealand Labour Party/New Zealand Worker Printing & Publishing Co.

Probine, M. and Fargher, R. (March 1987). *The management, funding and organisation of continuing education and training: Report of a ministerial working party.* Wellington: Authors.

Ramsay, P.D.K. (1985). The domestication of teachers: A case of social control. In J. Codd, R. Harker, and R. Nash (eds.), *Political issues in New Zealand education* (pp. 103–22). Palmerston North: Dunmore Press.

Raynor, J. (1972). The curriculum in England. In J. Raynor and N. Grant (eds.), *Educational studies: A second level course. The curriculum: Context, design and development units 2 and 3—patterns of curriculum* (pp. 5–71). Buckinghamshire: Open University Press.

Simon, B. (1969). Comprehensive schools. In E. Blishen (ed.), *Blond's encyclopaedia of education* (pp. 155–58). London: Blond Educational.

Step in right direction. (August 2005). *The Australian.* Reproduced in *The New Zealand Herald.* [Editorial]. (August 25), p. A14.

Tate, F. (1925). *Investigation into certain aspects of post-primary education in New Zealand* (Special report on educational subjects No. 16). Wellington: Government Printer.

Thom, A.H. (1950). *The district high schools of New Zealand.* Wellington: New Zealand Council for Educational Research.

Thomson, A. (October 26, 2005). Schools told: Help strugglers. *The New Zealand Herald*, p. A1.

Walker, R. (1985). Cultural domination of Taha Maori: The potential for radical transformation. In J. Codd, R. Harker, and R. Nash (eds.), *Political issues in New Zealand education* (pp. 73–82). Palmerston North: Dunmore Press.

Whitehead, C. (1974). The Thomas report—A study in educational reform. *New Zealand Journal of Educational Studies* 9 (1): 52–64.

Chapter 11

"My Parents Came Here with Nothing and They Wanted Us to Achieve": Italian Australians and School Success

Pavla Miller

In Australia, as in other countries, the unfinished project of making high schools "comprehensive" is under attack. In the 1970s, comprehensive schools were depicted as flawed but perfectible engines of democratic citizenship, neighborliness, equality of opportunity, and effective preparation for a diversity of productive lives. Today, national governments encourage market competition between schools, provide substantial subsidies to the private education sector, are skeptical about the worth of public institutions, and miserly with funding them.[1] The competition for enrolments is a serious business for government schools. In the last 15 years, many have closed despite strong community support because their enrolments fell beyond what education departments saw as sustainable. In turn, and often reluctantly, parents in many regions and social groups are coming to see comprehensive schools as residual institutions for those unable or unwilling to send their children elsewhere. There is compelling evidence that these developments are actively and systematically contributing to social polarization. Yet, as in the past, schools do not simply "reproduce" patterns of privilege and subordination in the wider society. Some regions, schools, programs, and social groups do considerably better than expected, others do worse. To provide a sense of this uneven process, this chapter sets research on inequality and educational outcomes alongside the trajectory of one relatively successful working-class group. Drawing on the accounts of family life by three generations of Italian Australians in Melbourne,[2] the chapter emphasizes the value of both systemic and "local" analyses.

After two decades of expansion and consolidation, Australia has one of the strongest private educational sectors among Organization for Economic Cooperation and Development (OECD) countries: only Netherlands, Belgium, and Spain have larger proportion of junior secondary enrolments in private schools. Importantly, both private and government contributions to the private education sector have outstripped rates of its enrolment growth. In 2000, the average share of private school enrolments at the junior secondary level among OECD countries was 16.4 percent. In Australia, at 30.9 percent, it was almost twice as much. Between 1984 and 2002, the *proportion* of all students enrolled in Australian government schools fell from 75 percent to 68 percent, the share in independent schools increased from 6 to 12 percent, and that in Catholic schools remained stable at around 20 percent. During this period, *enrolments* in government schools increased by less than 1 percent, the Catholic sector grew by 16 percent and the independent sector by 108 percent (Lamb, Long, and Baldwin, 2004, pp. 16, 40–41). Changes in enrolment patterns were compounded by redirection of funding. Between 1991 and 2000 there was a 2.7 percent increase in private school enrolments as a percentage of all enrolments, but a 4.6 percent increase in government funding for private schools. While the direction of *state* government funding has changed little, per capita *Commonwealth* funding per student in private schools increased by 94 percent between 1991 and 2000; that of per student in state schools grew by 56 percent. Among OECD countries, Australia also has relatively high—and increasing—rate of *private* expenditure on school education (exceeded only by Germany, Mexico, and Korea). Between 1991 and 2000, while *public* funding of school education increased by 60.4 percent, *private* funding—largely from parents of private school students—grew by 95.7 percent (Lamb, Long, and Baldwin, 2004, pp. 12–16).

In Australia, the late 1960 and 1970s are now considered as the high point of comprehensive high schools and of centrally controlled and bureaucratically managed public education more generally. University fees had been abolished, a national Schools Commission funded a number of innovative equity and affirmative action programs, and state schools, with around 80 percent of all enrolments, catered to a broad cross-section of the community. The progressive reforms introduced during this period drew on—and were subject to—a vigorous critique of schooling.

Australian scholars were among those who theorized the ways the "competitive academic curriculum" helps perpetuate class privilege by converting social background into school results. One of the most influential projects centered around the book *Making the Difference*, first published in 1982. Drawing on the work of Bourdieu, Bernstein, and others, Connell and his colleagues argued that the forms of solidarity, workplace orientation, and useful knowledge shared by working people constituted a handicap at school. In contrast, the cultural attributes engendered in managers' and professionals' lives were recognized as conceptual sophistication. Corporate and private schools, which could be seen as "organic" parts of the ruling class, played a powerful role in this process. State schools, particularly those in working-class suburbs, had a more tenuous relationship with the families they dealt with. Constrained by university requirements, and frequently despite the best efforts of their staff, they inflicted what could be called "hidden injuries of class" (Sennett and Cobb, 1972) on many of the children they taught. Among these injuries was the children's conviction that they "didn't have the brains" to do any better. Equality of opportunity, in sum, worked better for some groups, in some schools and regions, than others. In many demonstrable ways, working-class students and neighborhoods, indigenous Australians, non-English-speaking immigrants, and girls were disadvantaged. The way forward, articulated in a series of influential public reports, was to make schools and their curricula more comprehensive and inclusive.

The recent work of Teese and his colleagues from Melbourne University falls within the same broad tradition, but is written in a considerably different climate. Working in a period of retreat rather than consolidation of comprehensive schools, these scholars focus on the ways government policies and reforms are making schools *less* comprehensive and democratic. Combining sophisticated, system-wide statistical analyses with powerful theorization of school effects, the authors present compelling evidence that ongoing increases in expenditure targeted at private school students are making large and systematic contributions to social polarization. To some extent, this is because of the sheer disparity in resources between wealthy private schools and ordinary state schools. But the main effect, they argue, is through the differential composition of student groups. Here, the effects of residential segregation are amplified by the capacity of the wealthiest private schools to skim talented and motivated students from their "competitors."

Teese and Polesel (2003) argue that young people prefer curriculum areas where they work in groups, and where results are *tangible* and valuable in themselves. This is where their efforts have concrete results, which give them confidence in their own ability to transform the world. But doing well in these subjects, however valuable educationally, does not confer a competitive advantage in examinations for university entrance. Subjects such as English, chemistry, mathematics, and physics, in contrast, provide *intrinsic* satisfaction to those that succeed in them, rely on *individual* effort and responsibility, and revolve around a high degree of abstraction. In the Australian education system, they are also the most profitable subjects, facilitating entry to faculties preparing students for the most lucrative professions. The minority who do well in such subjects tend to enjoy them; to many, mathematics classrooms resemble a "laboratory where you learn to use tools," or even "a sports ground where you are training to win." Those for whom such subjects are impenetrable tend to experience the most prestigious parts of school curricula as boring, and see them in terms of "a factory where you are there to work" or an "office where you learn to follow routine." Teese and Polesel (2003, p. 104) note that "learning becomes work" for those who least benefit from its outcomes; "the graph of enjoyment exactly reversed the graph of meaninglessness." While many succeed, they conclude, it is their success relative to those on the other side of town that is significant.

Such arguments are similar to earlier work. What is novel is the authors' theorization, based on system-wide statistical analyses of exam results, of the compounding effects of polarized and increasingly competitive education markets. The result, Teese and Polesel show, is the development of schools where material and cultural resources and talent are concentrated, and others from which they are drained. By creating a rich pool of cultural and financial resources at selected sites, parents are able to pedagogically multiply individual advantages. The privileged schools tend to have smaller class sizes, better-trained teachers, more books, computers, and other resources, and students whose social background is more easily converted into examination success. Such pooling of resources in private schools "narrows the gap between results for the average child from a background of high socio-economic status and the individual child from such a background." (2003, p. 121) This works because pedagogical efficiency can be used where new or modest wealth was not accompanied by cultural capital, and also because the challenge of exams could be met collectively, rather than by individuals acting in isolation. In the process, the schools export failure:

> Private schools, operating on an assured platform of public grants, drain secondary education of the cultural resources represented by family education life-style and know-how and pump these into the most profitable locations of the curriculum. (Teese, 2000, p. 204)

On the other side of town are schools that lack resources, attended by students who pool disadvantage. These increasingly "residual" schools, Teese and Polesel note, cope with poor language skills, fragmented family lives, poverty, low levels of parental education, lack of facilities, high teacher turnover, and leisure that is distracting rather than supportive of school. Such schools are hard to staff; they cannot expel disruptive and demanding students, and see many of their most promising pupils transfer out to—or outright poached by—private schools. In one environment, the curriculum has been domesticated; in the other, it presents a continual real threat; here, many students see the school as a prison, and want to leave to get a job as soon as they can. Especially among those who get poor results, only a minority agrees that there are good careers for those who study hard, and few girls and fewer boys believe that extra efforts with schoolwork are worthwhile. As other educationist point out, it is by resisting such "boring" work that many children cement their failure in increasingly competitive education markets. As Willis (1977) argued in *Learning to Labour*, working-class lads earn self-respect, and the admiration of their peers, by resisting schoolwork in the most imaginative ways they can contrive. Such resistance is fueled by particular notions of sexist masculinity, where physical prowess is admired and bookish learning, pen pushing, and obedience to teachers is despised. In a cruel irony, the lads help produce their class subordination through successful resistance to subordination at school. Similar themes, this time from the perspective of working-class girls, have been developed by feminists writers such as McRobbie (1978, 2000) and Hey (1997).

Teese (2000, p. 2) provides graphic statistical evidence of the results of this process. In the urban regions of Melbourne where working-class and immigrant families are highly concentrated, every third girl can expect to receive failing grades in the least demanding mathematics subject in the curriculum. Among boys failure strikes more than 40 percent. The better students gravitate to the mathematics subjects that lead to university—but again, one in three fail. This is the same rate of failure experienced by preparatory mathematics students as a group in 1960. To attend school in the western suburbs of Melbourne, Teese concludes, is to be more than three decades behind in relative terms. To be educated in the upmarket inner east is to leave history behind, for here only 12 percent of boys and 8 percent of girls fail in preparatory mathematics. Similarly, in the dilapidated state schools in the working-class, reindustrializing western suburbs of Melbourne with their many immigrant communities, half or more of all boys fail English. In wealthy private schools, failure in English among boys comes in decimal points or zeros, with 1.2 per 100 an exceptionally high figure (Teese, 2000, pp. 208–09).

At the upper end of the examination register and in areas of the curriculum that count most toward university entry—mathematics and sciences, languages, traditional humanities—high marks again advantage the private system by a large margin. In chemistry, some 30 percent of private school students in Melbourne are placed in the top fifth band of achievement, compared with only 15 percent of public high school candidates. In physics, the margin is 29 percent against 16 percent; in biology 44 percent as against 15 percent; and in specialist mathematics, 28 percent compared with 15 percent. The more culturally selective nature of the students taking subjects such as Australian History and English Literature in private schools is generally associated with even larger relative advantages in performance: private school students are three times as likely as high school students to receive high grades, with every second candidate or more being placed in the top fifth band (Teese, 2000, p. 204). Finally, even though they dominate less prestigious universities and technical colleges, and fewer proportionately complete high school, students from state schools in the western suburbs of Melbourne are three times more likely than private school students to receive no offer at all of a tertiary place at the end of their high school study (24 percent compared to 7 percent) (Teese, 2000, p. 211).

Reviewing international comparative data on learning outcomes across a range of different skill areas, Teese and his colleagues note that, on average, the results Australian school children achieve are good by international standards. However, differences in achievement based on SES and other social factors are not only larger in Australia than in many OECD countries, but widen as young people progress through school. The authors conclude that, as a result of a relentless process of segregation, children do not have access to the same resources and learning opportunities. Those who enter private schools are more likely to enroll in the academic track. Those in academic tracks within senior high school tend to take subjects that are most profitable in terms of promoting *access to* university and prestigious courses *within* university. In all, the amplifying power of dissimilar student groups and their different relationship to curriculum areas results in substantial differences in school success and labor market opportunities. While non-Catholic private schools achieve, on average, stronger results in university entrance examinations, Catholic schools are associated with better school-to-work transition rates, as well as with higher rates of participation in further education and training (Lamb, Rumberger, Jesseon, and Teese, 2004, p. 45).

* * *

The main emphasis of the work summarized earlier is on system-wide effects of educational policies and practices. In effect, Teese and his colleagues document a process of making Australian schools less and less comprehensive. In the process, social privilege and disadvantage become amplified. What then are we to make of findings which show that some *social groups* with modest resources and little cultural capital disproportionately succeed at school? Italian Australians are a case in point. Nancy, a 45-year-old mother of two teenage children, was born in Melbourne, Australia to parents from the Abruzzo region of southern Italy. In an interview about families and children I recently conducted, she explained a successful strategy underpinned by hard work, family solidarity, and school success:

> I have been married for 25 years and . . . my parents have . . . always been there. . . . They've been supportive. They had expectations that we went to uni and that we studied because my parents came here with nothing and they wanted us to achieve . . . . rock mountains . . . and do all the kinds of things . . . I had different ideas . . . I wanted to . . . go to uni but I wasn't keen on doing that at 18 . . . and I wanted everything . . . to be married, have kids, I wanted to have a degree, I wanted to have a career, travel . . . And I did all those things . . . And my family helped me financially. I bought my first

home with my parents' help and in-laws' help. I married an Italian from the same part of Italy. So a lot of things worked out very well for me . . . I guess for me there's an expectation—my own—that I'm going to help my children to buy their house . . . but I do agree that kids have to learn responsibility and they have to understand the value of money . . .

Nancy completed a computer course and now works as an office manager. Her parents identify as Italian, she herself feels Australian, and her children claim they are Italian.

In terms of qualifications, many of Nancy's contemporaries fared less well than she did. Yet recent studies of the comparative standing of different ethnic groups in Australia (Long, Carpenter, and Hayden, 1999) and other English-speaking countries (Zhou, 1997; Waldinger and Perlman, 1998; Boyd and Grieco, 1998) show that children of many poor, badly educated, and exploited immigrants disproportionately succeed at school. Despite coming from lower SES background, a higher proportion of first-generation youth born in Asian or southern and eastern European countries complete secondary school and a larger share also continue on to university compared with their Anglo counterparts. A recent national investigation of the Australian-born children of postwar immigrants (Khoo et al., 2002) is a case in point. Based on reanalysis of the 1986, 1991, and 1996 Australian Census, the study made it possible to compare the educational outcomes of several different age cohorts of second-generation Australians. For Italians, the results showed that an earlier pattern of investing in sons' but not daughters' schooling more recently gave way to strong emphasis on the educational qualifications of both genders.

According to the 1996 Census, 23.2 percent of men aged 35–44 with both parents born in Italy had a degree or diploma, slightly *more* than third-generation Australians (at 22.5 percent). Another 29 percent of both groups had vocational qualifications. For women, the situation was reversed: 26.5 percent of third-generation Australian women had degrees or diplomas, compared to 22.4 percent of those with an Italian father, and only 18.8 percent of those with both parents born in Italy (Khoo et al., 2002, p. 81). Among *both* women and men who attended school a decade or two later, having both parents (rather than just the father) born in Italy now became a marked educational advantage. Among those aged 22–24, 20 percent of second-generation young men and 33.8 percent of women held a degree or a diploma, as opposed to 16.4 percent and 27.9 percent, respectively, for third-generation Australians. Conversely, about 51.5 percent of second-generation Italians had *no* post school qualifications, considerably fewer than the 57.1 percent of third-generation men and 60.7 percent of women (Khoo et al., 2002, p. 78). There was also far less class-based disparity in educational *participation* of 20–21-year-old second-generation Italian Australians than among their third-generation counterparts. Among third-generation men and women, 30.8 percent sons and 40.1 percent daughters and of managerial/professional fathers but only 6.7 percent sons and 11.1 percent daughters of those working in "other (mainly unskilled) occupations" attended university. Among the sons and daughters of Italian-born fathers, the disparity was far smaller: 25.9 percent sons and 36.4 percent of daughters of professional/managerial fathers were university students, compared to 18.2 percent of sons and 21.3 percent of daughters of those in "other occupations" (Khoo et al., 2002, pp. 67, 117). In other words, the ratio of what could be called class-based educational inequality between third-generation people was about four to one, among second-generation Italians, considerably less than two to one.

The combined effects of class, ethnicity, and school sector clearly complicate the links between family background and school success. In explaining these figures, three factors stand out: Italian's disproportionate capacity to save and invest in projects they value, the changing fortunes of the Catholic school sector in Australia, and the ability of some ethnic groups to achieve valued outcomes by *resisting assimilation*.

Escaping postwar shortages, unemployment, and poverty, most Italians came to Australia with "nothing but their suitcases" to secure a better life for themselves and their children. What they lacked in book learning they made up in shared understandings of the hardship and rewards of establishing a household, or *"systemazzione"* in distant lands and unfamiliar environments, and assembling a livelihood from a patchwork of low-wage and subsistence activities. Nancy's successful housing strategy is typical of her compatriots. Among those born in Australia of Italian parents between 1952 and 1961, 83.4 percent were living in their own home in 1986, 85.1 percent in 1991, and 87.5 percent in 1996. The corresponding figures for third-generation Australians were considerably lower: 63.4, 69.7, and 74.4 percent, respectively (Khoo et al., 2002, p. 78). The Italians' capacity to live frugally and to save, some unionist noted, made them more resilient when a strike was called. Having more savings and being "more clever with money," they could undertake more protracted industrial disputes. As two Italian workers at a large Ford plant in an industrial suburb of Melbourne explained to labor historians (Lever-Tracy and Quinlan, 1988, p. 304), "Australians can't sustain a long strike because they spend each week's wages before they get it. We learned during the war how to survive without food or clothes." "Australians are quick to go on strike but then they want to go back. You don't win that way."

The same capacity to secure a family's future could be mobilized in other ways. Explaining why people like Nancy's children, who often "have to buy nothing" when they marry, decide to have smaller families than their parents, 54-year-old Sandra noted:

> It's not that people don't want to have children because it doesn't please them to have children; it is too expensive to have children. You need to send them to [private] school. You need to dress them. You need to give them sport,

exercise, restaurants, all of these things that they want. And if I go to work and I don't have enough money to give them all of these things? And so, this is what I think. I have five children, but if I had to have children today, I'd only have two. Maximum three—but no more.

Education assumes particular significance in these calculations. Those who cannot afford private school fees, many grandchildren of Italian immigrants argue, simply should not have kids. As 21-year-old Amanda put it, "You'd want to be able to give them opportunities in terms of school, and ra ra ra. If you can't afford it then . . . that's my view anyway. I'd like to send my kid to [private] school and try and give 'em a head start sort of thing.'" A friend added "Yeah, that's pretty much a general kind of view, isn't it?" "No conflicts there," she concluded, after everyone in the group of young adults agreed. In conversations such as these, the assumption that children need to attend *private* schools is so widely shared that it is not stated.

In their encounters with Australian schools as parents and students in the 1950s and 1960s, most of the Italian Australians I interviewed recalled minor humiliations punctuated by occasions of intense conflict. Those who were able to "pass" pretended to be Anglo-Australian to avoid widespread racism. Pressured by the local Catholic clergy to send children to parish schools, Italian families were less likely than their Australian neighbors to patronize state schools. This hardly constituted a worldly advantage. If anything, the schools' Catholicism was Irish, with little tolerance for non-English speakers. While a few boys' schools run by Catholic brothers and priests established a reputation for success in public examinations, provision for girls focused on home making, nursing, and teaching (O'Donoghue, 2001). Until the 1960s, Catholic schools received no state aid. Catering to many of the new European immigrants, their enrolments rose sharply in the postwar decades. Although costs were contained with large classes and low pay of teaching congregations, by the late 1960s the schools faced a crisis: the necessity to employ increasing numbers of lay teachers put most in considerable financial difficulties. When the national Karmel Committee reported on schools throughout Australia in 1973, it identified Catholic schools among those in greatest need and catering to the greatest proportion of disadvantaged students. The Labor Commonwealth Government's adoption of the Committee's recommendation to provide needs-based funding for *all* Australian schools saved many from closure, and allowed the restoration and expansion of the Catholic system of low-fee secondary schools. Today, the increasing flow of resources toward the private school sector in Australia has made attendance of most (not all) Catholic schools a relative advantage rather than disadvantage.

The majority of Catholic schools, catering to a socially more diverse body of students, lack the cultural exclusivity that fuels the pedagogic power of wealthy independent schools. In a scattergram plotting school achievement[3] in the final year of high school against SES in 1998 (Teese and

Polesel, 2003, p. 120), almost all independent schools lie within the first quadrant (high SES and high results), and many are concentrated high up in that quadrant. Catholic schools are spread across quadrants 1 (high SES and high results), 3 (low SES, low results), and 4 (low SES, high results), but tend to be more heavily concentrated around the center. Government high schools also display a large spread, but only a few schools are at the high end of quadrant 1, while a large number are "deep" into quadrant 3.

Whether or not they have accurate appreciation of the comparative benefits of particular schools, Italians are among those leading the flight from the government to nongovernment education sector. In a valuable project on the history of comprehensive high schools in New South Wales (NSW), Campbell and Sherington document some of the dimensions of this process. Stressing that national school policies have different effects in different regions, and for different class and ethnic groups, the authors note that by 2000, only 38 percent of youth of Southern European ancestry in NSW attended state schools, as opposed to 66 percent for all Australians (Campbell and Sherington, 2006, p. 141). Of all the groups surveyed, Italians had by far the weakest attachment to state schools and the strongest attachment to Catholic schools. In the inner west region of Sydney, with its strong and long-established system of Catholic schools, children born of Italian mothers were evenly split between state and Catholic schools in 1976. Twenty years later, the few remaining state comprehensive schools only enroled 21 percent of youth with Italian-born fathers. The vast majority, 76 percent, attended Catholic schools, and another 3 percent attended other nongovernment schools (Campbell, 2003, p. 590; Campbell and Sherington, 2006, pp. 144–45). Catering to 76 percent of Italian students, Catholic secondary schools were the dominant provider, at a share that approached the highest level of state school enrolments for all children in the post–World War II period. Children of Greek mothers, in contrast, had a far greater—but still declining—attachment to state schools. In 1976, 94 percent of Greek children in this part of Sydney were enrolled in state schools; by 1996, the proportion was 66 percent.

In the aforementioned account, Italians' greater propensity to save has been used as part of an explanation for their modest but still disproportionate educational success rates. Such considerations of subaltern ethnicity as a strength rather than a problem have recently received particular emphasis in sociological analyses. Connell, like Willis and, initially at least, many of his feminist critics, focused on the dynamics of class and gender but not ethnicity or race. In the radical education literature of the 1970s and 1980s, working-class immigrants, Aborigines, and people of color were frequently designated as "triply disadvantaged," handicapped at once by poverty, lack of familiarity with school knowledge and customs, sexism and racism; their disadvantage compounded by attendance of impoverished schools. Since then, a generation of scholars and activists developed compelling critiques of the ethnocentrism, racism—and empirical shortcomings—of what can be called a deficit approach to ethnicity. Today, it is routinely acknowledged that "ethnic families," despite their poverty, can be a source of strength and resistance in a hostile Anglo society.

Among the most coherent theorization of these issues are accounts of "segmented assimilation," used by a number of North American authors to explain apparently anomalous findings regarding the educational attainment of children of immigrants. For many decades, Portes and Zhou (1993) argue, assimilation to the (generally unexamined) norms and values of American society was represented as a good thing. Today, a considerable body of evidence shows that for some groups, the successful *avoidance* of assimilation, and maintenance of immigrant community cohesion, is a superior path to health and prosperity. Not only do less "assimilated" immigrants enjoy better health status; their children succeed at school against great odds, and at a far greater rate than their more "assimilated" peers. The reasons, according to Portes and Zhou, are relatively straightforward. In impoverished city neighborhoods, decimated by deindustrialization, anti-egalitarian public policies and racism, young people of color preserve their dignity by developing a vibrant oppositional culture. Like Willis's lads, they oppose school; they also buy and sell drugs, are frequent victims—and perpetrators—of violence, have little access to good jobs—and believe, reasonably enough, that there is no point in slaving at school. Newly arrived immigrants who "assimilate" into this oppositional culture become American but fail at school; those who for one reason or another maintain the hopes of their parents and observe the moral norms of a tightly knit ethnic enclave tend to succeed. They devote more time and energy to schoolwork, not least since allegiance to the cultural project of their ethnic communities helps "switch off" resistance to school. Finally, ethnic communities such as the Italians, who for generations perfected strategies for the acquisition of economic security through the pooling of concerted effort of all family members, are well placed to invest in education markets.

Such positive possible effects of ethnicity on educational outcomes, however, are blocked for children of "socially defined racial minority groups." Mexicans, Chicanos, Dominicans, and Haitians in the United States and Aboriginal children in Australia might well wish to assimilate but find that normal paths of integration are blocked on the basis of race. Such groups devise alternative strategies of coping with racial barriers. Since they perceive the effects of the education system to be continued exploitation, such strategies are hardly likely to encourage school success (Zhou, 1997, pp. 986–89).

Conclusion

The retreat from comprehensive high schools, this chapter argued, has gone further in Australia than in most other OECD countries. In analyzing the impact of this process, two groups of Australian scholars have developed approaches that can be of use in other parts of the world. Using sophisticated social theory and extensive survey data, Teese and his colleagues not only provide nuanced descriptions of cultural processes through which education markets compound social polarization, but are able to provide unusually detailed, geographically specific, and statistically innovative measures of comparative educational outcomes. In focusing on regional peculiarities of the same broad processes, Campbell and Sherington show conclusively that education systems cannot be adequately studied on the national level alone. For a range of historical, systemic, and accidental reasons, different regions and social groups negotiate the current restructuring of secondary schools in diverse and often unexpected ways.

Insights such as these can be set within recent debates on the politics of geographical scale by social demographers and welfare state theorists (Delaney and Leitner, 1997; Mahon, 2005, 2006). Taking as read the premise that economies, policies, and social actors operate within complex hierarchies of space, these scholars have begun to examine the ways in which *notions of scale* are socially constructed, and in turn have tangible social effects. Much of the literature deals with the way welfare states, with their entrenched scalar configurations, are being reorganized, rejigged, and retrenched in the current era of global, national, and local restructuring (Brenner, 2001, p. 592). In Australian education, a particularly significant aspect of this process concerns the highly contested selection of statistical indicators for allocating "needs-based" funding. As Teese (2004) argues, the Commonwealth government applies an "SES methodology" to funding nongovernment schools, which is insensitive to real need and which disproportionately benefits wealthy schools. This is because it bases definition of need on the average income level of the statistical districts where individual children reside. This is inequitable, Teese argues, because it does not take into consideration "creaming" school intake policies, and assumes that children enrolled in wealthy independent schools are typical of the immediate locality in which they live.

In education more generally, there are significant, contested, and changing configurations of actors, capacities, and histories at international, national, and local levels. Regional factors continue to be important; the localities in which schools are built and students reside differ markedly in the extent of residential segregation and social polarization, economic vitality and wealth. Yet families can and do relocate to other countries, send students to schools in distant suburbs or cities, enroll children in studies for the International Baccalaureate. In turn, contested architectures of school provision, curricula, and teaching approaches not only select out different social attributes as marks of intelligence, they assemble quite different mixes of students inside classrooms. In this way, they constitute significantly different filters for converting social background, financial resources, assiduity, and talent into examination results. It is in international projects such as this book that the dimensions of this process become more clearly visible.

Notes

1. Under the Australian constitution, the funding of government schools is a state responsibility, but federal governments play an increasing role through per capita subsidies to schools. Both Labor and non-Labor parties are committed to "state aid," with Labor governments supporting greater needs-based funding and Liberal governments being more generous to elite private schools.
2. The project, funded by a small ARC grant in 2000, involved interviews with 50 grandparents, parents, and children broadly representative of postwar Italian immigrants in Melbourne.
3. General Achievement Test administered in year 12 in Victoria in 1998.

References

Boyd, M. and Grieco, E. (1998). Triumphant transitions: Socioeconomic achievements of the second generation in Canada. *International Migration Review* 32 (4): 853–76.

Brenner, N. (2001). The limits to scale? Methodological reflections on scalar structuration. *Progress in Human Geography* 25 (4): 591–614.

Campbell, C. (2003). State policy and regional diversity in the provision of secondary education for the youth of Sydney, 1960–2001. *History of Education* 32 (5): 577–94.

Campbell, C. and Sherington, G. (2006). *The comprehensive public high school: Historical perspectives*. New York: Palgrave Macmillan.

Connell, R.W. et al. (1982). *Making the difference: Schools, families and social division*. Sydney: Allen and Unwin.

Delaney, D. and Leitner, H. (1997). The political construction of scale. *Political Geography* 16 (2): 93–97.

Hey, V. (1997). *The company she keeps: An ethnography of girls' friendships*. Buckingham: Open University Press.

Karmel, P. (Chair) (1973). *Schools in Australia: Report of the interim committee of the Australian schools commission*. Canberra: AGPS.

Khoo, S.E., McDonald, P., Giorgas, D., and Birrell, B. (2002). *Second generation Australians*. Canberra: Australian Centre for Population Research and the Department of Immigration and Multicultural and Indigenous Affairs, AGPS.

Lamb, S., Long, M., and Baldwin, G. (2004). *Performance of the Australian education and training system*. Report for the Victorian Department of Premier and Cabinet, Centre for Post-Compulsory Education and Lifelong Learning, University of Melbourne.

Lamb, S., Rumberger, D.J., and Teese, R. (2004). *School performance in Australia: Results from analyses of school effectiveness*. Report for the Victorian Department of Premier and Cabinet, Centre for Post-Compulsory Education and Lifelong Learning, University of Melbourne.

Lever-Tracy, C. and Quinlan, M. (1988). *A divided working class: Ethnic segmentation and industrial conflict in Australia*. London: Routledge & Kegan Paul.

Long, M., Carpenter, P., and Hayden, M. (1999). *Longitudinal surveys of Australian youth, Participation in education and training, 1980–1984*. Melbourne: Australian Council for Educational Research.

Mahon, R. (2005). Rescaling social reproduction: Childcare in Toronto/Canada and Stockholm/Sweden. *International Journal of Urban and Regional Research* 29 (2): 341–57.

——— (2006). Introduction: Gender and the politics of scale. *Social Politics* 13 (4): 457–61.

McRobbie, A. (1978). Working class girls and the culture of femininity. In Women's Studies Group, Centre for Contemporary Cultural Studies, University of Birmingham. *Women Take Issue*. London: Hutchinson.

——— (2000). *Feminism and culture* (2nd ed.). London: Macmillan.

O'Donoghue, T. (2001). *Upholding the faith: The process of education in Catholic schools in Australia 1922–1965*. New York: Peter Lang.

Portes, A. and Zhou, M. (1993). The new second generation: Segmented assimilation and its variants among post-1995 immigrant youth. *Annals of the American Academy of Political and Social Sciences* 530: 74–98.

Sennett, R. and Cobb, J. (1972). *The hidden injuries of class*. New York: Knopf.

Teese, R. (2000). *Academic success and social power*. Melbourne: Melbourne University Press.

——— (2004). Submission to senate employment, Workplace relations and education references committee, inquiry into commonwealth funding for schools.

Teese, R. and Polesel, J. (2003). *Undemocratic schooling: Equity and quality in mass education in Australia*. Melbourne: Melbourne University Press.

Waldinger, R. and Perlman, J. (1998). Second generation: Past, present and future. *Journal of Ethnic and Migration Studies* 24 (1): 5–24.

Willis, P. (1977). *Learning to labour: How working class kids get working class jobs*. Westmead: Saxon House.

Zhou, M. (1997). Segmented assimilation: Issues, controversies and recent new research on the new second generation. *International Migration Review* 31 (4): 975–1008.

Epilogue

Chapter 12

Epilogue—The Future of the Comprehensive High School

Gary McCulloch and Barry M. Franklin

This volume of essays has attempted to broach a number of issues around the development of the comprehensive high schools that have troubled observers in America and around the world in recent years. Contributors to the volume have located a range of problems afflicting this major institution, and also some ways in which it has attempted to respond to these difficult challenges. This has been attempted with the intention avowed by Muller, Ringer, and Simon (1987) two decades ago, of beginning to move beyond "*ad hoc* descriptions of particular institutions, and of strictly national narratives," toward "socio-historical and comparative approaches to educational *systems*" (Muller, Ringer, and Simon, p. 1). We should like at this stage to review the patterns identified and the conclusions that have been reached so far and also to assess some lines of future research to build on these foundations.

Half a century ago, James Bryant Conant (1959) made the grand claim that "the American high school has become an institution which has no counterpart in any other country" (p. 7). This great and characteristic American invention was the comprehensive high school, which unlike specialized high schools provided a general education for all future citizens, elective programs for those who wished to use their acquired skills immediately on graduation, and also programs for those whose vocations would depend on their subsequent education in a college or university. Conant went on to argue:

> Though generalization about American public education is highly dangerous . . . I believe it accurate to state that a high school accommodating all the

youth of a community is typical of American public education. I think it safe to say that the comprehensive high school is characteristic of our society and further that it has come into being because of our economic history and our devotion to the ideals of equality of opportunity and equality of status. (p. 8)

At least three of the statements in this account have come into question in the present volume: the extent to which the comprehensive high school accommodated all the youth of a community, the notion that it was characteristic of American society, and the idea that it had no counterpart in any other country.

Our contributors have suggested a more varied picture, socially and regionally, within the United States, and a number of stresses and strains that have created pressure on the institution. Some of these have been internal to the schools, while others have grown externally from the changing features of American society. These have also been replicated, to a greater or lesser extent, in the comprehensive high schools that have developed in other countries.

They have also shown that other developed countries in different parts of the world have been strongly influenced by this basic prototype or template. In a number of countries, the comprehensive high school has been installed as a key feature of secondary education at different stages of the past century. On the other hand this does not necessarily support the view of evolutionary progress proposed by Benn and Simon in Britain in the early 1970s. According to Benn and Simon (1970) at that point, the basic comprehensive model was winning the battle against other types of secondary school, including in western Europe, as "part of a world wide movement concerned to adapt the structure of secondary education to the new demands of scientific, technological, social, racial and cultural progress" (p. 36). Their argument was maintained, and even developed further, in a follow-up study also in Britain a generation later with the subtitle "Is comprehensive education alive and well or struggling to survive?" As the authors of this volume noted, "There may well be individual exceptions that prove the rule, but in general most countries start off with highly selective systems and gradually translate to comprehensive ones as they move up the industrializing scale, particularly if they wish to compete at the top" (Benn and Chitty, 1996, p. 20). Yet different nations and regions appear again to have developed in a more complex and varied way than this view would suggest. In the United States, for example, the degree of selectivity, at least compared to other systems of state schooling was quite minimal and lasted for a shorter period of time. Moreover, the social and political challenges of the past decade have made it less than clear whether this one best system will survive unscathed, much less that it will prosper and dominate the scene across the board.

Nevertheless, it would be premature to conclude that the comprehensive high school is dead or dying. Reports of its demise have indeed been greatly exaggerated. The uncertainties of purpose and identity, the intimations of mortality that have been much in evidence throughout this volume might be read as signs of a mid-life crisis rather than of the final death-rattle. Moreover, as we have also seen, in some areas there are indications of robustness and rude health, and indeed of fresh growth.

It is fitting in some respects that the United States with its long history of comprehensive secondary education may also be the site for its rebirth and ultimate survival. As some of the contributors to this volume have suggested, the institutional form that the first comprehensive high schools took often did not fulfill the institution's basic promises. And similarly, the core values of the comprehensive high schools, universality and inclusiveness, can be attained through other structures that perhaps reduce the institution's size, alter its regulative mechanisms, and provide for diverse delivery schemes. The end result may not be the demise or death of the comprehensive high school but its partial transformation into any of a number of new forms that are more authentic in attaining its core purposes.

Clearly the comprehensive high school faces a combination of difficult issues and challenges, but it is not likely to go quietly into that good night. In one important sense, its decline and demise would be a "strange death" along the lines of that of Liberal England (Dangerfield, 1935/1997). The travails of the institution are in marked contrast to the continued resonance of the principles that it is held to embody, in particular equality of opportunity and democracy, something that a number of our contributors have stressed in their essays. Perhaps, as with many other great American institutions, its difficulties herald not death, but reinvention and a spectacular comeback.

Finally, in keeping with our view that there is still a great deal of work to be done in this field, may we suggest a few of the needs and opportunities for scholars to respond to in the next few years. There is a need for greater detailed recognition of the role of different social and ethnic groups, and of a wide range of different case studies of institutions and localities, to enhance our understanding of the position of the comprehensive high school. Similarly, there is much scope for work on the everyday practices of the schools, involving the daily routines of the curriculum, the classroom and school grounds, teachers and pupils. Clearly we have only begun the long road to embracing the wide range of national and regional examples that should be developed in their broader international and global context. The position and prospects of comprehensive high schools in developing nations in different parts of the world require further investigation. There is also more to say about their relationships with the state in different contexts, although it is important to strike out into new directions with fresh assessments of their dealings with private schools and to understanding the drivers of selection and inequality in a supposedly nonselective and equal system. We should see much more research that explores the processes of globalization and migration and their implications for these schools. All of this research, we would argue, would benefit from being based in an understanding of the interrelationships of historical, contemporary, and comparative concerns, developed through combinations of research strategies that highlight individual experience and agency at the same time that they seek out macro-systems and structures. None of this will pick at the bones of the comprehensive high schools; it should rather help us to bring them to life.

References

Benn, C. and Chitty, C. (1996). *Thirty years on: Is comprehensive education alive and well or struggling to survive?* Harmondsworth: Penguin.
Benn, C. and Simon, B. (1970). *Half way there.* Harmondsworth: Penguin.
Conant, J.B. (1959). *The American high school today: A first report to interested citizens.* New York: McGraw-Hill.
Dangerfield, G. (1935/1997). *The strange death of liberal England.* Stanford: Stanford University Press.
Muller, D., Ringer, F., and Simon, B. (Eds.) (1987). *The rise of the modern educational system.* Cambridge: Cambridge University Press.

Contributors

Rene Antrop González is Assistant Professor of Curriculum and Instruction/ Second Language Education, University of Wisconsin-Milwaukee.

David Crook is Senior Lecturer in History of Education and Assistant Dean of Research and Consultancy at the Institute of Education, University of London.

Anthony De Jesús is Assistant Professor of Social Work, Hunter College School of Social Work, City University of New York.

Barry M. Franklin is Professor of Secondary Education and Adjunct Professor of History at Utah State University.

Gregory Lee is Associate Professor of Policy, Cultural, and Social Studies at the University of Waikato (New Zealand).

Howard Lee is Professor of Educational Studies at Massey University (New Zealand).

Gary McCulloch is Brian Simon Professor of the History of Education and Dean of Research and Consultancy at the Institute of Education, University of London.

Pavla Miller is Professor of Historical Sociology at the Royal Melbourne Institute of Technology University (Australia).

Roger Openshaw is Professor of Educational Studies at Massey University (New Zealand).

Thomas C. Pedroni is Assistant Professor of Social Studies Education and Policy Studies at Oakland University.

Jośe R. Rosario is Professor of Educational Foundations and Curriculum and Instruction at Indiana University-Purdue University, Indianapolis.

John L. Rury is Professor of Education at the University of Kansas.

Sevan G. Terzian is Associate Professor and Program Coordinator of Social Foundations of Education at the University of Florida.

Wayne J. Urban is Associate Director of the Education Policy Center and Professor of Educational Leadership, Policy, and Technology Studies at the University of Alabama.

Susanne Wiborg is Senior Lecturer in Comparative Education and Life Long Learning at the Institute of Education, University of London.

Index

ability, 23, 24, 43, 138, 150, 154
  grouping by, 25–6, 42, 132
Aborigines, 194, 195
Abruzzo, 190
academic approach, 7, 38–9, 41, 42, 43, 77, 120, 140, 143, 151, 160, 170, 180, 190
academic expectations, 10, 74, 75, 78, 84, 85
academy school, 147, 154, 164
accountability, 9, 154, 176
Achieve Inc., 9–10
adscription, 136
adult education/schooling, 59, 60, 62
affirmative action, 186
Africa, 83
African-Americans, 11, 54, 60, 62, 65, 79–80, 83, 85, 88, 91, 104, 111–27
after-school activities, 79
Aguila, Ramona, 123
Albizu Campos, Dr Pedro, 84
algebra, 83
American Association of Junior Colleges, 41
American Association of School Administrators (AASA), 37, 39, 40, 50
American City, 43–5, 47
Anderson, C., 97
Anglesey, 149, 150, 157
Anglo-Australian, 193
Angus, David, 179–80
Antrop-Gonzales, Rene, 10, 205
  article by, 73–92

Aorere College, 178
Appiah, Kwame Anthony, 93–4
apprenticeship, 162
aristocracy, 135, 137
Arredondo, P., 86
art, 25, 81, 82, 83, 154, 173
arts, 44, 147, 178
Aspen Institute, 10
aspirations, 24
At Risk Youth, 121
Attlee, Clement, 149
Auckland, 5, 178
Australia, 4, 5, 13–14, 185–98
authoritarianism, 96, 97–100

baby boom, 67, 169
Baccalaureate, International, 196
Bayliss, Trevor, 156
Beard, Charles A., 39, 50
Bedfordshire, 159
Beeby, Dr C.E., 171–4, 175, 178
Belgium, 186
Benn, Caroline, 163, 202
Benson, T., 97
Berkeley, 40
Bernstein, Basil, 186
bilingual/biculturalism, 74, 76
Bill and Melinda Gates Foundation, 10
biology, 48, 82, 83, 189
Birmingham (UK), 5
Bismarck, Otto von, 133
Bjerregård, Ritt, 143
Black Alliance for Educational Options, 111
Black Papers, 152, 162

black school, 96, 113–14
  black controlled private schools, 115
black students/population, 9, 53, 57, 59, 60, 62, 76, 103, 114–15
Blair, Tony, 13, 148, 155–6, 157, 161, 162
Blunkett, David, 154
Board of Education, 79, 148
Bondeförbundare, 141
Bondeförbundet, 141
bottanskola, 141
Bourdieu, Pierre, 186
bourgeoisie, *see* class, social
Bowles, Samuel, 84
Boyle, Edward, 159
Boynton, Frank David, 22–3
Boyson, Rhodes, 153, 162
Bradley Foundation, 115
Britain, 4, 5, 133, 136, 174, 202
  see also, England, Wales
British Broadcasting Corporation (BBC), 156, 157
British Liberal Party, 136
  see also, Liberal Party/Parties *and* liberalism
Brown (vs Board of Education of Topeka, 1954), 111–12
Buckinghamshire, 153
business, 27
busing, 114–15
Butt, Ronald, 153

calculus, 83
Callaghan, James, 153
Cambridge University, 102
Campbell, C., 5, 194, 196
Canada, 148, 196
Cardinal Principles of Secondary Education (1918), 20, 25, 38, 113
care/caring, 10, 75–8
Caribbean, 82
Carnegie Institution, 47
Carr, William, G., 40, 41
case study, 78
Catholic Center Party, 133

Catholic influences/schools, 117–18, 120, 161, 175, 186, 190, 192, 193, 194
CCC, The NYA, and the Public Schools, The, 46
census, 67
Census, Australian, 191
Census, U.S. Bureau of the, 67
Central Park East Secondary School, 10, 95–6, 100–5, 106–8
chemistry, 25, 83, 187, 189
Chicago, 10, 74, 80, 81, 84
Chicago Tribune, 80
Chicanos, 195
Chitty, Clyde, 163
choice, 28–9, 117, 125, 143, 153, 154, 158
choir, 105
Christensen, I.C., 138
Circular 10/65, 149, 151, 152, 159
Circular 10/70, 152
Circular 10/74, 152
citizenship, 20, 32, 34, 43, 44, 61, 102, 113, 136, 185, 201
city technology college, 153
civic preparation, 7
civil disobedience, 83
Civilian Conservation Corps (CCC), 19, 40, 42, 46, 47
Civil Rights (Movement), 65, 66, 95, 112–15
civil society, 107
class, social, 11, 32–3, 43, 66, 84, 85, 91, 111, 113, 116, 125–6, 132, 134, 135, 136, 137, 185, 188, 189, 196
  bourgeoisie, 137, 139
  collaboration, 139
  middle-class, 74, 136, 150, 152, 155, 158
  mixed social backgrounds, 138
  peasantry, 143
  privilege, 186
  working class housing, 150, 187
  working class opportunity, 149

classical languages, 25, 138
class size, 118–27, 149, 188
clerical employment, 63, 64, 67
Cleveland, 74, 112
clubs, 20, 32–3, 34
Cold War, 45
college, 27, 118, 120–1
color, communities of, 75, 77
color, people/students of, 12, 74, 75, 112
color, schools of, 73–91
Comintern, 140, 143
commercial subjects, 25, 27, 30
Commission on the Reorganization of Secondary Education (1918), 8
Committee of Ten (1894), 8
common learnings, 41
communitarianism, 100, 105
community, 44, 45, 54, 55, 73, 74, 75–7, 79, 80, 83, 96, 107, 185
  activism, 83, 86
  of color, 81, 90, 115, 116
  control, 75
  garden, 82
  geographical, 10–11
  government through, 95, 102–4
  Latina/o, 75, 77, 80, 83, 85, 86, 91
  learning community, 10, 79, 85, 87
  school as, 100–1
  schools within, 12, 34, 43
composition, 29
comprehensive high school
  challenges to, 3–4, 66, 202
  criticisms, 7, 9–10, 34, 118–24, 125–6, 155–6, 178
  death of, 4, 13, 73, 90, 111, 147–8, 203
  decline of, 153–4
  and democracy, 20–1, 34, 42, 43, 44, 53, 96, 98, 102, 185, 203
  and economy, 23–4
  expansion and growth, 6, 8–9, 23, 53, 202, 203
  failure, 74
  future, 14, 143, 164, 203–4
  meanings of, 157–61
  origins, 131–43, 148–52, 170–4
  preparatory role, 27–8, 31–4, 43–4, 94, 185
  purposes, 7, 19, 53–4
  rationale, 3
  reform, 8, 11, 37–8, 66
  successes and failures, 161–4
  vocational preparation, 30–2, 43
comprehensive ideal, 3, 53–4, 58, 60, 61, 64, 66, 113, 176, 178
computer lab/course, 81, 191
Conant, James Bryant, 39, 201–2
Congress, 46
Connell, R.W., 186, 194
consensus, 139, 140, 143
Conservative Party, 153, 162
conservatives, 135, 137, 149, 152
Consultative Committee on the Post-Primary School Curriculum (1942), *see* Thomas Committee
consumerism, 44
contagion model, 55–6
Core Curriculum Review (1984), 176
Cornell University, 20, 21, 22, 28, 29, 31
counseling, 28–9, 30–1
Court of Appeal, 153
crafts, 173
creaming, 196
crime, 122–3
Crook, David, 13, 205
  article by, 147–67
Crosland, Susan, 151
Crosland, Tony, 149, 151, 152, 159
Cuba, 82, 84
cultural capital, 188, 190
culture, 79, 82, 83, 84, 85, 86, 89, 119, 122, 125, 134, 136, 150, 171, 202
curriculum, 9, 22–7, 29, 30, 32–4, 41–2, 43, 44, 47–50, 73, 75, 77, 80–6, 89, 118, 126, 160, 169, 172, 179, 187, 188, 203
  academic, 19, 20

curriculum—*continued*
  breadth, 158
  common, 14, 170, 172, 173, 174, 176
  differentiated, 6–7, 11, 22–3, 125, 170
  Eurocentric, 80
  hidden, 84
  integrated, 81–2
  liberalization, 174
  national, 163, 176
  Sankofa, 81–2
Curriculum Review (1984, 1986), 176
Currie Commission (1962), 175
Cusick, P., 96
custodialism, 7–8, 19–20, 26–7, 33–4

Daily Mail, 147
dance, 82
Danish Liberal Party, 135
  see also, Liberal Party/Parties *and* liberalism
Dapedako, Dasha, 118–19
Dartmouth College, 30
Davis, Evan, 157
deficit, 195
De Jesús, Anthony, 10, 205
  article by, 73–92
Denmark, 131–9, 142–3
Department for Education and Science, 151
Department of Education, 174
desegregation, 112, 113, 114–15
design, 154
Detroit, 74
diploma, 26, 27, 191
discipline, 12, 29, 120, 122, 126, 163
discrimination, 32
diversity, 117–18, 153, 154, 158, 178
Division Street Business Development Association (DSBDA), 81
documentary research, 78
Dominican Republic, 82
Dominicans, 195
Domino Sugar factory, 81–2

Donoughue, Bernard, 160
Dr. Pedro Albizu Campos Alternative High School (PACHS), 10, 74–5, 77–8, 80–1
dropouts, 27, 54, 64, 86, 91, 97
Duane, Michael, 163
Dworkin, R., 94
Dyke, Greg, 156

East Riding of Yorkshire, 158
Education Act, 1944, 148
Education Act, 1976, 153
Educational Policies Commission (EPC), 37–50
Educational Reconstruction (1943), 148
Education and Skills Committee, 157
Education for All American Youth (1944), 8, 37, 42–50
Education in Change (1969), 175
Education Today and Tomorrow (1944), 173
egalitarianism, 96, 100, 104, 137, 155, 195
Eight Year Study (Progressive Education Association), 8
elementary school/education, 24, 131, 132, 138, 141
eleven plus (11 plus), 148, 150, 152, 154, 156–7
elitism, 32, 124, 134, 136, 137
El Puente Academy for Peace and Justice, 10, 74–5, 77–8, 79–80
Elvis Costello, 167
employment, 9
engineering, 25, 147
England, 12, 13, 147–67, 179
English, 25, 26, 27, 29, 41, 46, 47, 81, 82, 173, 176, 187, 189
enrollments, 8–9, 10, 53–67, 185
entrepreneurial approach, 177
equality, 112, 142, 162
equality of opportunity, 3, 7, 148, 164, 171–2, 174, 176, 178, 185, 187, 201, 202
Esping-Andersen, Gösta, 139

Essential Grammar School (1956), 150
essential schools, 101
Ethics of Identity (2005), 93–4
ethnicity, 11, 55, 74, 80, 83, 118, 191, 192, 194, 195, 203
ethnography, 78, 112, 116
Eton College, 159
Europe, 132, 133, 136, 137, 155, 202
examinations, 6, 143, 149, 175, 187
experiential learning, 82, 119
extra-curricular activities, 20, 32–3, 34, 41, 97

Failed Promise of the American High School, 179–80
faith schools, 164
family, 66, 74, 86–9, 112, 113, 115–24, 126, 187, 190–5
Farmville, 43–5, 47
federal government, 43, 46, 47
feminism, 75, 77
fine arts, 81
focus groups, 78
Folketing, 135
foreign languages, 25, 26, 27, 147, 154, 178, 189
Forum for the Discussion of New Trends in Education, 163
Forum for Promoting 3–19 Comprehensive Education, 163
Foucault, Michel, 94, 106
foundation school, 154, 164
France, 12
Franklin, Barry M., 205
  articles by, 3–16, 201–4
Fraser, Peter, 171, 172, 175
Freire, Paulo, 87
Friedman, Milton, 112
Frisinnade, 141
Frontiers of Democracy, 46
Fuller, Howard, 111, 112, 125
funding of education, 11, 23–4, 43, 45, 47, 78, 122, 124, 125, 140, 141, 154, 169, 177, 185, 186, 193, 196, 197

further education, 131

Gaitskell, Hugh, 149–50, 160
Gallardo-Cooper, M., 86
Garden-Acosta, Luis, 79
Gardner, Howard, 118
Gayford, Keith, 179
gender, 76, 97, 98, 123, 171, 187, 189
General Achievement Test, 197
General Educational Board, 39, 40
gentrification, 81, 83
geometry, 83
Germany, 12, 39, 133, 139, 186
Gintis, Herbert, 84
global studies, 81, 82
Gordon, C., 106
governance, 9–10, 94–5, 102–3
governmentality, 94–5, 97, 102–3, 106–8
graduation rates, 9
grammar, 29
grammar schools, 12, 131, 148–54, 155–6, 158, 159, 160, 161, 162, 175
Grant, B.K., 95, 96
grant maintained school, 154
Great Depression, 7, 19, 20, 22, 23, 24, 26, 33, 39
Green, Andy, 5
Greenberg, J.B., 81
Grundtvig, N.F.S., 138, 144
Grundtvig folk high schools, 136
Guam, 84
guidance, 20, 25, 29, 30, 33, 44
guidance department, 23, 27, 28, 30–3
gymnasium, 12, 132, 133

Hadow Report, 179
Hague, William, 162
Hailsham, Lord, 150
Haitians, 195
Hammack, Floyd, 34
Hanan, Josiah, 170
Harlem, 10
Hartford, 84

Harvard Redbook, General Education in a Free Society (1945), 8, 48–9
Harvard University, 38, 39, 48
Hauptschule, 133
Hawaii, 84
health education, 47–8, 198
Heath, Edward, 152
Herder, J.G., 144
heritage, 76
Hey, V., 189
higher education, 12, 24, 27, 30, 31, 37, 38–9, 118, 120–1, 160, 162, 177, 187, 189, 190, 191, 201
Higher Standards, Better Schools for All (2005), 155
High School (1968), 10, 93–100, 106
High School II (1992), 10, 93–6, 100–6
history, 25, 31, 41, 81, 138, 189
HIV/AIDS, 81
Hogben-Seddon free place system, 170
Holland Park Comprehensive School, 159
homework, 120
Horsbrugh, Florence, 149
Hounslow Secondary Modern, 157
House of Commons, 153
House of Lords, 153, 159
humanities, 101, 189
Humboldt Park, 81
Hutchins, Robert, 42

ideology, 49, 96, 97, 98, 117, 136, 139, 142, 155, 159
immigration, 55, 185–98, 193
industrial arts, 25, 27, 48
inequality, 6, 8–9, 65, 66, 185, 203
information technology, 154
Inspectorate, 149
intellectuals, 42, 141, 151
intelligence, 43, 196
intelligence tests, 26
interdisciplinarity, 41
international comparisons, 9
interviews, 78, 116–17, 125
Irish, 193

Islington Green School, 160
Israel, Sonia, 116
Italian Australians, 14, 185–98
Italy, 190, 191
Ithaca, New York State, 21
Ithaca High School, 7–8, 20–34

Jackson, P.W., 95
James, Eric, 162
Japan, 39
Jenkins, Simon, 162
Jensen, Ole Vig, 143
Johnson, Gary, 122–3
Johnson, Lyndon Baines, 97
Jørgensen, Jørgen, 142
journalism, 83
June Constitution, 135
junior high school, 24

Karmel Committee, 193
Kemble, Bruce, 160
Kennedy, Bobby, 97
Kent, 153
Kidbrooke, 150
Kidbrooke School, 163
kindergarten, 102
King, Martin Luther, 97
knowledge economy, 9
Knowledge Ventures Learning Academy, 121–4
Kohanga Reo, 177
Korea, 186
Kulp, Claude, 23–6, 29, 31, 33
Kura, Kaupapa, 177

Labor Commonwealth Government, 193, 197
Labor Democrats (Arbeiderdemokraterne), 144
labor market, 19, 23, 26, 55
Labour Party, 148, 149, 151, 153, 154, 155, 159, 160, 164
Lake View High School, 80
languages, *see* foreign languages
läroverk, 132

Latin, 25, 160
Latin America, 82, 91
Latina/os, 73, 76–7, 78, 79, 80, 84, 88, 89, 90, 124
Latina/o communities, 75, 77, 80, 83, 85, 86, 91
Latino schools, 10
Learnfare, 121, 127
learning community, 10, 79, 85, 87
Learning to Labour (1978), 189
Learning to Live (1958), 149
Lee, Gregory, 14, 205
   article by, 169–8
Lee, Howard, 14, 205
   article by, 169–8
Leicestershire, 158
Leicestershire Plan, 159
left, the, 46
leisure, 44
Liberal Party/parties, 13, 133–7, 138, 139, 141, 142, 143, 197
liberals/liberalism, 37, 133–7, 139
life history, 116
linear school system, 133
Lincoln, Jan, 120–1
linguistic alienation, 74
Listowel, Earl of, 149
literacy, 136
local control, 46
local education authority (LEA), 148–9, 150, 151, 152, 153, 154, 158, 161
London, 148, 149, 156, 157, 158, 163
London County Council, 149, 163
London Oratory School, 157, 161
London School Plan (1947), 158
Los Angeles, 95, 105
Lovett, Trevor, 157
Luebbert, G.M., 134
lycee, 12

Maden, Margaret, 160
magnet school, 114
Major, John, 153, 159
Making the Difference (1982), 186

Manchester, 149
Manchester Grammar School, 158
Mangere (Primary) Principals' Association, 179
Maori people, 176, 177
Mariama Abdullah School, 116
markets, educational, 11, 111, 112, 116–17, 125, 185
Marshall, Russell, 176
masculinity, 189
Mason, Rex, 173
mathematics, 25, 26, 27, 101, 173, 176, 178, 187, 189
McCarthy, Eugene, 97
McCulloch, Gary, 155, 205
   articles by, 3–16, 201–4
McKamey, C., 77, 90
McLaren, Ian, 180
McManus, Declan, 157
McRobbie, A., 189
medical school, 30
Meier, Debbie, 104
Melbourne, 14, 186, 189, 190, 192
Melbourne University, 187
Merelman, Richard, M., 93
metal shop, 25
Mexicans, 80, 83, 195
Mexico, 186
Meyer, John, 54, 55, 57, 65, 67
Michigan, 105
middle class, *see* class, social
middle school, 101, 131–2, 133, 136, 137–8, 142, 151, 159
militarism, 97
military service, 39
Miller, Pavla, 14, 205
   article by, 185–98
Milner, Frank, 171, 172
Milwaukee, 74, 112, 113–15, 116–18, 121, 122
Milwaukee Area Technical College, 119, 121
Milwaukee Parental Choice Program, 11, 116–18, 123
Ministry of Education, 149

Mirel, Jeffrey, 179–80
mixed ability class, 131, 132, 142–3, 155, 156, 160
Moll, L.C., 81
monarchy, 137
Muldoon, R.D., 177
Muller, D., 201
Mulley, Fred, 153, 160
multiple intelligences, 118
multiple regression, 9, 58
Muñoz Marín, Luis, 84
Murphy, Samantha, 121–2
music, 25, 31, 173
    teacher, 163

National Board of Youth Service (fictional), 43
National Commission on Excellence in Education (1983), 8
National Education Association (NEA), 8, 37, 38, 40, 46, 50, 51
National Government, 176, 177, 178
National Governors Association, 9–10
National Liberals, 135, 138
    see also, Liberal Party/Parties and liberalism
National Youth Administration (NYA), 19, 40, 42, 46, 47
A Nation at Risk (1983), 66
neoliberal(ism), 117, 143
Netherlands, 186
New Deal, 40
New Education Fellowship (NEF), 171
New South Wales, 5, 194
New York City, 10, 74, 79–80, 95, 105
New York Daily News, 90
New York State, 7, 21, 23, 26
New York Times, 79
New Zealand, 4, 5, 13–14, 169–84
New Zealand Curriculum Framework (NZCF), 178
New Zealand Herald, 179
New Zealand Labour Party, 171
    Labour Government, 173

New Zealand Post-Primary Teachers Association (NZPPTA), 175–6
New Zealand Secondary Schools Association, 172
Noddings, Nel, 76
non-academic subjects, 26
Norfolk, 159
Normal Entrance, 25
Northamptonshire, 159
North Division High School, 115
Northeast (USA), 55, 56, 57, 59, 60, 63
Northeast High School, 95–100, 106–7
Northern Ireland, 151
Northumberland, 159
Norway, 131–41, 142
Norwegian Social Democratic Party, 140, 143
    see also, Social Democratic Party/Parties
Norwood Report, 173, 179
numeracy, 178

Onehunga High School, 178
Openshaw, Roger, 14, 180, 205
    article by, 169–8
orchestra, 150
Orfield, Gary, 57
Organization for Economic Co-operation and Development (OECD), 5, 14, 143, 162, 186, 189, 195
overcrowding, 24, 73, 119, 123, 125
Oxford University, 102, 162

Pacific states (USA), 57, 60
parallel school system, 131, 141
parents, 22–3, 28–9, 34, 38, 43, 73, 80, 87, 99, 104, 111, 115–17, 118–24, 126, 143, 150, 151, 154, 158, 185, 188, 194
parliamentary sovereignty, 134
Parr, Christopher, J., 170
participant observation, 78

part-time employment, 63, 64, 119
paternalism, 98
paying students, 46–7
Pearl Harbor, 39
peasantry, 136–7, 138, 143
pedagogy, 74, 77, 80, 82, 89, 90, 101, 107, 118
  critical, 87
Pedroni, Thomas, C., 11–12, 205
  article by, 111–27
Pennsylvania, 10, 95
performativity, 154
performing arts, 154
Philadelphia, 10, 95, 97, 106
Philippines, 84
Phillips, Melanie, 147, 162
photography, 83
physical education, 25, 47, 173, 178
physics, 31, 187, 189
Picot Committee/Report, 177–8, 179
Pimlico Comprehensive School, 158
placement, 27, 28, 30
poetry, 81, 138
polarization, 14
Polesel, J., 187
policy (makers), 3, 8, 37–52, 62, 64, 66, 159, 161
polytechnics, 177
Popkewitz, T.S., 94
Portes, A., 195
Portland, 74
postcode lottery, 154
postmodernity, 155
poverty, 9, 47, 63, 64, 66, 97–100, 188
power, 87, 97
Prentice, Reginald, 161
Prescott, John, 156
Price, Gina, 119–20
primary education, 131, 163, 177, 180
Probine-Fargher Report (1987), 177
professions/professional employment, 63, 64, 67
proficiency examination, 170, 172
Program for International Student Assessment (PISA), 5, 143
progressive educational reform, 46
Progressive Education Association, 8
progressivism, 8, 179–80
private education, 14, 111–27, 158, 160, 185–90
private secondary schools, 11, 38, 132, 187, 203
private tutors, 154
Prussia, 135
psychology, 43
psychometric tests, 148, 150
Puerto Rican Cultural Center (PRCC), 80
Puerto Rican experience/scholars, 74–5, 90
Puerto Ricans, 74–5, 79–81, 83
Puerto Rico, 82, 84
pupils, *see* students

Quigley, Derek, 177

Realschule, 133
race/racism, 11, 55, 64, 66, 74, 83, 103, 114, 118, 122, 125, 193, 194, 195, 202
racial discrimination, 61
racial integration, 57
racial segregation, 10–12, 57
Radical Liberals (Radikale Venstre), 135, 142, 143
  see also, Liberal Party/Parties *and* liberalism
Ramsay, P.D.K., 177
Rée, Harry, 150, 151, 158, 163
regional differences, 13–14, 54–5, 56–66, 196, 202–3
religion, 117–18, 120, 133, 136, 154, 161, 175, 186, 190, 192, 193, 194
Renwick, W.L., 174
Report of the Taskforce to Review Educational Administration (1988), 169
Retsforbundet, 142
Riddle, Oscar, 47

right, the, 13, 46, 47, 135, 137
Ringer, F., 201
Ripon, 154
Risinghill Comprehensive, 163
Rodney King (trial), 95, 105
Rodríguez de Tió, Lola, 84
Roosevelt, Franklin, D., 40
Rosario, Jose, R., 10, 96, 205
　article by, 93–110
Rose, N., 94–5, 106–7
Ross, Nick, 156
Rotherham, 162
rural areas and urban/rural divide, 14, 43–5, 47, 133–9, 170
Rury, John, L., 8, 34, 205
　article by, 53–70
Ruskin College speech, 153
Russia, 39
Ryden, Värner, 141

St. John Stevas, Norman, 153, 159
Saint Urbina Catholic High School, 117–21
Santiago-Rivera, A., 86
Saunders, Chris, 178
Savage, Graham, 148
Scandinavia, 4, 12–13, 131–45
scholarship ladder, 149
school leaving age, 44–5, 47, 180
School Review, 49
Schools Commission, 186
School Standards and Framework Act, 154
science, 25, 26, 27, 46, 47–8, 81, 85, 101, 175, 176, 178, 189, 202
Science, 47–8
scientific understanding, 44
secondary modern, 148, 150, 152, 155, 157, 158
segregation, 5, 121, 190, 196
Seguín High School, 76–7
selection, 5, 6, 12, 14, 131, 138–9, 142–3, 147, 152, 154, 156, 158, 159, 160, 162, 169, 170, 171, 174, 179, 202, 203

self-discipline, 29–30
Self-Help (1859), 136
setting, 142, 155, 160
sex, 97
sexism, 194
Shaw, Sandi, 156
Sheerman, Barry, 157
Shepherd, Gillian, 159
Sherington, G., 5, 194, 196
Short, Ted, 152
shorthand, 31
Simon, Brian, 150, 157, 163, 201, 202
sixth form, 158, 175
Sizer, Theodore, 101
skills, 21, 44, 83, 126, 178, 188, 201
Skoglund, Phil, 175
slavery, 82
small schools, 9–10, 74, 90, 101, 118–19, 125–6
Smiles, Samuel, 136
Snow, C.P., 159
social capital, 55
Social Democratic Party/Parties, 13, 134, 136, 139, 140, 141–3, 144
socialism, 139
socialists, 148, 149, 157, 159
social justice, 74, 85
social mobility, 4, 136
social studies, 25, 26, 101, 176, 178
sociology, 176
South, the (USA), 9, 56, 57, 58, 59, 60, 61, 64, 65
South America, 83
Southeast (USA), 63
Spain, 186
Spanish, 74
specialist school, 147, 154, 164
Specialist Schools and Academies Trust, 164
specialization, 4, 5, 13, 113, 153, 154, 158, 179, 201
spelling, 29, 156
Spens Report (1938), 148, 173, 179
sport, 147, 154
Squire, Robin, 159

Index

Staff Government, 135
stakeholders, 73, 112
standardization, 11, 126
standards, 7, 19, 155, 160, 161, 169, 175
Star Spangled Banner, 84
state, the, 45, 106–7, 136, 137, 139, 141, 186, 197, 203
Stewart, Michael, 151
streaming, 132, 142, 160
students, 7, 27–34, 39, 73, 77–8, 81–90, 95, 114, 118–24, 203
    passivity of, 97
subtractive schooling, 74, 79
Subtractive Schooling (1999), 76
suburbs/suburban, 54, 59, 196
Sugar Project, 81–2
Sutton, 159
Sveidrup, Johan, 137
Sweden, 131–9, 141–2
Sweet Freedom Sugar Feast, 82
Sydney, 194

Taíno people, 82
Tameside, 153
Tattler, 32–3
Taylor, Sir Cyril, 164
teachers, 7, 26, 28, 31, 33, 38, 39, 75–6, 78, 119, 120, 122, 151, 159, 160, 176, 203
    as facilitators, 87
    veteran, 124
teacher training, 9
technical school, 14, 140, 148, 162
technology, 65, 154, 175, 176, 178, 202
teenage parents, 121
Teese, R., 187–90, 196
temperance, 141
Terzian, Sevan G., 7–8, 205
    article by, 19–36
testing, 26, 143, 154
Thatcher, Margaret, 152, 153, 159, 164
Thomas Committee/Report (*Consultative Committee on the Post-Primary School Curriculum*, 1942), 169, 172–4, 178
Thompson, A., 76–7
Times, 153, 162
Times Educational Supplement, 163
Tomkins County, 21
Tomorrow's Schools (1988), 169, 177–8
tracking, 114, 140, 143
Trafford, 153
transformatory education, 77, 82–3, 84, 85, 91, 187
trigonometry, 83
trust school, 154
Tuley High School, 80
typing, 31

under-achievement, 101
unemployment, 40, 107
uniform, 120
United States of America (USA), 3–12, 19–128, 148, 179–80, 195, 201–4
universal suffrage, 135, 139
university, see higher education
Urban, Wayne, J., 8, 206
    article by, 37–52
urban areas/issues/rural divide, 14, 43–5, 47, 73, 80, 134, 135
US Office of Education, 8

Valenzuela, Angela, 76–7, 79, 89, 91
values, 96
VIDA/SIDA, 81
video, 82, 83
Vieques, 84
Vietnam War, 95, 97, 99
violence, 105
vocational study, 27, 29, 30, 46, 47, 49, 53, 61, 121, 180
vocational training, 8, 19, 21, 22–3, 24, 27–8, 45, 170, 171, 172
    and employers, 22
vouchers, 11–12, 111–12, 116–17, 118, 124, 125

Waitaki Boys' High School, 171
Wales, 13, 147–67
Walker, Patrick Gordon, 152
Walker, R., 177
Washington D.C., 112
Watford Grammar School, 150
Welfare Reform Act, 127
Wellington, Mary, 176
Wells High School
West, the (USA), 55, 56, 59, 60, 63
West Riding of Yorkshire, 149
white-collar (employment), 55, 57, 59, 60, 61, 67, 139, 141
white flight, 126
Whitty, Geoff, 164
Wiborg, Susanne, 5, 12, 206
   article by, 131–45
Wilkinson, Ellen, 149
Williams, Mike, 178
Williams, Shirley, 153, 155
Willis, Paul, 189, 194, 195

Wilson, A.N., 155, 156
Wilson, Harold, 151, 152, 160
Wisconsin, 112
Wisconsin Works, 127
Wiseman, Frederic, 10, 93–108
Withernsea High School, 158
women, 175, 176
working class, *see* class, social
World Bank, 5–6
World War I, 38, 134, 138
World War II, 6, 7, 13, 19, 21, 33, 37, 39, 41, 53, 56, 57, 132, 133, 139, 140, 162, 164, 169, 194

York, 151
Youth Commission of the American Council on Education, 40

Zhou, M., 196
Zinn, Howard, 83